Scientific Style in English

RIVER PUBLISHERS SERIES IN INNOVATION AND CHANGE IN EDUCATION - CROSS-CULTURAL PERSPECTIVE

Indexing: All books published in this series are submitted to Thomson Reuters Book Citation Index (BkCI), CrossRef and to Google Scholar.

Nowadays, educational institutions are being challenged as professional competences and expertise become progressively more complex. This is mainly because problems are more technology-bounded, unstable and ill-defined with the involvement of various integrated issues. Solving these problems requires interdisciplinary knowledge, collaboration skills, and innovative thinking, among other competences. In order to facilitate students with the competences expected in their future professions, educational institutions worldwide are implementing innovations and changes in many respects.

This book series includes a list of research projects that document innovation and change in education. The topics range from organizational change, curriculum design and innovation, and pedagogy development to the role of teaching staff in the change process, students' performance in the areas of not only academic scores, but also learning processes and skills development such as problem solving creativity, communication, and quality issues, among others. An inter- or cross-cultural perspective is studied in this book series that includes three layers. First, research contexts in these books include different countries/regions with various educational traditions, systems, and societal backgrounds in a global context. Second, the impact of professional and institutional cultures such as language, engineering, medicine and health, and teachers' education are also taken into consideration in these research projects. The third layer incorporates individual beliefs, perceptions, identity development and skills development in the learning processes, and inter-personal interaction and communication within the cultural contexts in the first two layers.

We strongly encourage you as an expert within this field to contribute with your research and help create an international awareness of this scientific subject.

For a list of other books in this series, visit www.riverpublishers.com

Scientific Style in English

Marcelo Sampaio de Alencar

Institute of Advanced Studies in Communications (Iecom)
Federal University of Bahia (UFBA)
Brazil

Thiago Tavares de Alencar

Institute of Advanced Studies in Communications (Iecom)
Brazil

Routledge
Taylor & Francis Group
LONDON AND NEW YORK

Published 2017 by River Publishers
River Publishers
Alsbjergvej 10, 9260 Gistrup, Denmark
www.riverpublishers.com

Distributed exclusively by Routledge
4 Park Square, Milton Park, Abingdon, Oxon OX14 4RN
605 Third Avenue, New York, NY 10017, USA

Scientific Style in English / by Marcelo Sampaio de Alencar, Thiago Tavares de Alencar.

Routledge is an imprint of the Taylor & Francis Group, an informa business

ISBN 978-87-93609-28-0 (print)

While every effort is made to provide dependable information, the publisher, authors, and editors cannot be held responsible for any errors or omissions.

*This book is dedicated to Silvana, Raissa, Raphael,
Janaina, Marcella, Vicente and Cora.*

Contents

Preface

This book is about writing in a scientific style. It contains helpful tips and guidelines for reading and writing technical, scientific, and mathematical papers, articles and books in English.

It is recommended for students and professionals with at least a basic or intermediate understanding of the English language.

The main focus of the book is on technical texts, concentrating on style, grammar and math.

The book deals with the many possible problems, which can be encountered when writing articles, reports, submitting a book proposal, and many more stylistic, technical and grammar related issues.

Chapter 1 introduces the definition of style, teaches the reader about scientific writing, presents guidelines for writing articles, reports and books, and how to submit a book proposal. It also showcases the elements of scientific writing, numbers and symbols and recommends some useful reference material.

Chapter 2 covers the mathematical style in English, ranging from reading to writing, while also dealing with the combination of numbers, words, formulas and theorems.

Chapter 3 focuses on technical and scientific writing, including the process of preparation of a document, the organization of an article, and how to coordinate figures and tables. It deals with revision and style, word choice and usage, grammar and punctuation. Some basic problems in technical writing are discussed, along with useful writing tips.

The steps to obtain higher focusing capability, faster and better reading, and time management while also retaining the essential information in the most efficient manner, are discussed in Chapter 4. The chapter also gives tips on how to read fast and more efficiently.

Chapter 5 deals with stylistic problems in English, ranging from grammar, to syntax, style, technical issues, expressions, figures, symbols and tables. It also explains how to utilize these elements correctly and enhance readability, clarity and organization of the technical text.

Chapter 6 is a guide on the creation of a book. It presents a thorough list of items that comprise a general book, including in-depth explanations of the front matter, body text, and the back matter of the book.

Chapter 7 covers the subject of theses and dissertations, similar terms that represent the required submissions for a master's degree and doctorate. It includes guidelines and general recommendations for writing a thesis or dissertation, which are useful for students of any course.

The differences between oral communication and written communication are discussed in Chapter 8, which also gives tips on how to talk in from of an audience, and how to prepare a successful presentation.

Every chapter contains a commented list of reference material, which includes books, articles, reports and web-pages.

There is an appendix with assignments for the readers to practice the material of the book. The bibliography contains all the reference material, in alphabetical order, and the remissive index helps the reader to find most of the essential terms of the book.

Marcelo S. Alencar

Thiago T. Alencar

List of Abbreviations

Ω	Omega
A	Ampère
AINP	American Institute of Physics
ATM	Asynchronous Transfer Mode
ATSC	American Television System Committee
BGP	Border Gateway Protocol
ca.	Circa
CDF	Cumulative Distribution Function
cf.	Confer
e.g.	Exempli gratia
ESL	English as a Second Language
et al.	Et alii
etc.	Et cetera
et seq.	Et sequientes
ftp	File Transfer Protocol
f.v.	Folio verso
GIF	Graphics Interchange Format
H	Henry
HTTP	Hypertext Transfer Protocol
Hz	Hertz
ibid.	Ibidem
id.	Idem
IEEE	Institute of Electrical and Electronics Engineers
i.e.	Id est
IETF	Internet Engineering Task Force
inf.	Infra
IP	Internet Protocol
IPv6	Internet Protocol version 6
ISBN	International Standard Book Number
ISSN	International Standard Serial Number
LCD	Liquid Crystal Display

loc. cit.	Loco citato
n.b.	Nota bene
NIST	National Institute of Standards and Technology
OED	Oxford English Dictionary
op. cit.	Opere citato
OSPF	Open Shortest Path First
OWL	Online Writing Lab
Ph.D.	Philosophiæ Doctor
PSTN	Public Switched Telephone Network
Q.E.D.	Quod erat demonstrandum
q.v.	Quod vide
RIP	Routing Information Protocol
RSVP	Resource Reservation Protocol
RTP	Real-time Transport Protocol
SS7	Signalling System number 7
sup.	Supra
s.v.	Sub verbo
TPC	Transmission Control Protocol
T	Tesla
UDP	User Datagram Protocol
URL	Universal Resource Locator
ut sup.	Ut supra
viz.	Videlicet
vs.	Versus
v.	Versus
W	Watt

1

Scientific Style in English

"Most of the fundamental ideas of science are essentially simple, and may, as a rule, be expressed in a language comprehensible to everyone."
Albert Einstein

1.1 Introduction

Style is a distinctive manner of expression, in writing or speech. The word evolved from Latin *stilus*, an instrument for writing, marking, or incising, such as something used by the ancients in writing on clay or waxed tablets.

It is the way in which something is said or done, as distinguished from its substance. Style is also a convention with respect to spelling, punctuation, capitalization, and typographic arrangement and display followed in writing or printing.

This book aims to help with reading and writing scientific and technical texts in English, with focuses on style, grammar and math. It can be used by students and professionals with basic or intermediate understanding of the English language.

This chapter covers most of the topics in the book, introduces the notation and symbology, explains most of the terms, presents the general rules of style and contains a great deal of the book references, most of them commented.

1.2 Scientific Writing

Part of the charm of a good fiction book is the implicit nature of the sentences, which gives the readers a certain freedom to imagine. In this regard, literature is the art of imprecision. When an author says, in a romance, "I am falling for you," it does not mean that the character is actually being attracted by the Earth's gravity, in this passage of the book.

1.2.1 Good Scientific Writing

Scientific writing, differently from a poetry or a romance, must be accurate, concise, useful, clear, illustrated with visuals, targeted to a specific audience, well-organized, interesting, consistent, complete, correct in spelling, punctuation and grammar. Therefore, it is possible to list the following important points related to scientific literature:

- Scientific precision – Inaccurate statements destroy the author's credibility. On the other hand, the readers make decisions, operate equipment, and draw scientific conclusions based on the information presented.
- Concision – It is important to avoid wordiness caused by:
  - modifiers: *whole system, final outcome, completely finished*;
  - coordinated synonyms: *basic and simple, each and every, basic and fundamental*;
  - excess qualification: *perfectly clear, completely accurate*;
  - expletives, relative pronouns, and relative adjectives: *make it clear that ..., there are ..., who ..., which*;
  - circumlocution: *a long, indirect way to express the whole and complete idea.*

  The author must also avoid repetition, pompous language, jargon, and consider that reducing a document is hard work. Blaise Pascal once wrote, "I have made this a long letter because I haven't the time to make it shorter."
- The document must be useful – People read paper, technical report, or thesis because they intend to use the information in some way. Therefore, each sentence must contain useful information.
- Clearness – Keep the writing short and simple by breaking the text into concise sections, and avoid jargon or slang, because unknown terms cause poor communication and also make the text obscure to the readers. Present the story in a logical, orderly fashion, one step at a time. The use of visuals is recommended.
- Illustrate the manuscript with visuals – Visuals make the document more interesting to the reader. The author can use flowgraphs, photographs, drawings, diagrams, graphs, tables, and flowcharts.
  - **flowgraphs** (represents, using graph notation, the paths that might be traversed by a program during its execution).
  - **photographs** (show actual physical images of subjects).
  - **drawings** (depict real or imaginary objects, and internal parts).

- **diagrams** (show how the components interact and are interrelated).
- **graphs** (show trends and how one variable changes in relation to another).
- **tables** (organize information systematically).
- **flowcharts** (show the parts or steps in a process or system interact).

- The author should target an audience – Study the public first, and write to the level of technical proficiency and understanding of the audience.
- Organize the document – Plan, before writing, creating a rough outline that spells out the contents and organization of the document.
- Interest – The paper competes with many other communications and, therefore, must be lively and lucid, to attract the reader, not dull, repetitive and boring.
- Consistency – Inconsistencies confuse readers and convince them that the scientific work is as sloppy as the author prose. Avoid random and unnecessary capitalization, mixed sets of units of measurement and indiscriminate use of abbreviations.
- Completeness – A complete document tells the readers all they need to know about the topic, but not more. Make sure the specification is complete and that nothing important is omitted.
- Correction in spelling, punctuation and grammar – This is a key characteristic of every good paper or book. Equations, formulas and expressions also need to follow the rules of grammar, mainly punctuation.

1.3 Writing Process

A successful writing process involves several phases, which include preparation, research, organization, writing and revision. It is important to follow the steps and to adhere to the scope of the work.

Preparation

- Establish the purpose of the document – What should the readers know after they have finished reading the document?
- Assess the audience – Who exactly is the reader, or what are the interests of the readers? Who needs to see, or who wants to profit from the document? What the expected readers know about the subject?

- Consider the context – Context is the environment, or set of circumstances, in which writers produce documents and within which readers interpret their meanings.
- Determine the scope of the coverage – The decisions on what to include, and what not to include, in the writing define the scope of the document.
- Select the appropriate medium – Decide in what journal or conference to publish the results of the research, based on the areas covered by the publication, and on the intended public.

Research

Research involves the question: What does the researcher know about the subject? Then, the research (primary or secondary) can be conducted, creating and using questionnaires, if appropriate, and interviewing to obtain information. In the next phase the author must summarize the information, take notes, document the sources and, not to be forgotten, avoid plagiarism.

Organization

Organization involves the choice of the best methods to develop the work. Then it is necessary to outline notes and ideas, develop and integrate visuals. It it important to carefully consider the layout and design.

Writing

During the writing phase, select an appropriate point of view, and adopt an appropriate style and tone. Use an effective and positive sentence construction. Write effective paragraphs from the points on the outline.

Use quotations and paraphrasing, if appropriate for the document, always citing the sources of information, that should appear at the end of the document. After that, write an introduction for the document, then write an abstract, write a conclusion and choose a title.

Principles of Technical Communication

If possible, always use the active voice in the document and a plain, rather than a complex or elaborate language. Delete words, sentences and phrases that do not add to the meaning, and use specific and concrete terms rather than vague generalities.

Use terms that the intended reader can picture. Use the past tense to describe the experimental work and results, but, in most other writing, use the present tense. Break the writing into short sections to facilitate the perusal of the document.

Revision

Heraclitus (ca. 540–470 BC) was a Greek philosopher who stated that reason is the only constant in an ever-changing world. During the revision process, check for unity and coherence, verify sentence variety, emphasis, and subordination, check for ambiguity, awkwardness, and verify logical errors:

- lack of reason is when a statement is contrary to the reader's common sense.
- sweeping generalizations are statements too broad to be supportable by any theory.
- *non sequitur* is a statement that does not logically follow a previous one.
- false cause refers to the logical fallacy that one event followed another one, the first event caused the second.
- biased, or suppressed evidence, is both unreasonable and unethical.
- facts, which are logically or empirically verifiable, versus opinion.
- a loaded argument occurs when a conclusion is only based on an opinion.

Aim at positive writing, consider ethics, copyright, avoid biased language and plagiarism. Verify the appropriate word choice, abuse of affectation and jargon, do not use vague words (*real, nice, important, good, bad*), and clichés (*last but not least*).

Minimize the usual problems with grammar. Review punctuation, abbreviations, capitalization, contractions, dates, italics, number and measurement units, proofreading, and spelling.

1.4 Writing an Article

A document is a report, or a description, of reference material, usually written or in electronic format, for an article, a book, a thesis, a product, but also for sales invoices, wills and deeds, newspaper issues, newspaper stories, history recordings, executive orders, commercial letters, and product specifications.

An electronic document on a computer is created using a word processor or a text editor. A document usually adheres to some convention based on similar or previous documents or specified requirements.

1.4.1 Parts of the Document

- Title, authors and affiliations – Capitalize the initial letters of the words of the title, except articles (a, an, the), coordinating conjunctions (and, but), or prepositions (to, at, into) unless they begin or end the title.
- Abstract – Highlight and summarize, in 50 to 200 words, the major points of an article. This is the base for researchers to decide whether to read your article. It must be readable apart from the the original document.
- Keywords – Three to five words related to the main theme of the article.
- Introduction – Point out the purpose of the paper, define the problem examined, and present the scope of the article.
- Methodology – Discuss previous work in the field, the rationale of the approach to the problem and the reason to reject alternative approaches.
- Results – This section contains the main results of the research.
- Discussion – An optional section with a discussion of the main results, commonly used in Medicine and Biomedical articles.
- Conclusions – Usually the final section, it interprets the results in relation to the purpose of the study and the methods used to conduct it. The conclusions must depend entirely on the evidence found in the research. Comparisons with results found in the literature are welcome.
- References – It is necessary to cite the references in the body of the document, not far from the point in which they are mentioned, and list them at the end of the document.
- Appendix – Appendices are optional, and typically include explanatory material or a mathematical deduction.

1.5 Writing a Formal Report

A formal report is an official document that contains detailed information, research, and data necessary to make academic, industrial, technical or business decisions. The report is usually written for the purpose of explaining a technology, defining a problem, studying a case, or solving a problem.

Some examples of formal reports include an academic report, an inspection report, a safety report, a technical report, a compliance report, an audit, an incident report, an annual report, a situational report.

The formal reports are classified into two categories: informational and analytical reports. The informational report collects data and facts, that are used to draw conclusions. The analytical report contains similar information as the informational report, but it also offers recommendations to solve a certain problem.

A formal report must include the front section, the main section and the back section.

1.5.1 Front Section

The front section of a formal report is what the intended audience sees first. Therefore, the front cover must contain the report's title and the author's name. It should also contain the date of publication, but no page number. The title, name, and date should be evenly spaced to achieve an appropriate balance on the page. The title should be written in a larger font size than the name and date. Use initial capitals for the title.

- Title, authors, organization, target institution, and date – Capitalize the initial letters of the words of the title.
- Abstract – Summarizes the major points of the report. Must be readable apart from the the original document.
- Table of contents – List of main topics of the document.
- List of figures – All the figures must be in this list.
- List of tables – All the tables must be in this list.
- Foreword – An optional introductory statement that is written by, usually, an authority in the field or executive of the organization sponsoring the project.
- Preface – An explanation of the purpose of the document, and a description of its contents.
- List of abbreviations and symbols – Mandatory for theses and books, and optional for papers.

1.5.2 Main Section

The main section of the formal report contains an executive summary, an introduction, a discussion, a set of conclusions of the report, some recommendations and, eventually, explanatory notes.

- Executive summary – Provides a more complete overview of the report, stating its purpose, scope, and gives background information. It also summarizes conclusions and recommendations.
- Introduction – Points out the purpose and defines the scope of the report, and examines the problem.
- Text – Contains the details of the investigation, the alternatives explored, and visuals and tables to clarify the explanation.
- Conclusions – This section interprets the results and must depend entirely on the evidence found.
- Recommendations – If necessary, recommendations are included in the document.
- Explanatory notes – If necessary, notes that explain details of the work are included in the document.

1.5.3 Back Section

The back section portion of the report contains the appendices, glossary, and references. This type of material usually follows the conclusion.

- References – A list of works cited.
- Appendices – Contains additional material that can be useful to the reader to clarify or supplement the body.
- Bibliography – An alphabetical list of all sources that were consulted in researching the report.
- Glossary – An alphabetical list of specialized terms used in the report and their definitions.
- Index – An alphabetical list of specialized major topics discussed in the report, along with page numbers. The index is always the final section of the report.

1.6 Submitting a Book Proposal

A proposal is a document written to persuade someone to follow a plan of course to fulfill a need.

1.6.1 An Outline of a Book to Be Submitted to an Editor

A book proposal is addressed to the publishing editor or director and includes a title, the author name and affiliation, background information, a brief description of the book, the reasons for writing the book, the book market and readership, the competition, a table of contents, manuscript information, author information, a sample writing and the assessment process.

- Title – A clear and accurate title is important in marketing your book. As a general rule, the main title should have no more than seven words. If the title looks like it will be longer than that, then consider using a subtitle.
- Author name and affiliation – You should include your mailing address, e-mail, and phone and fax numbers.
- Background – This should outline the general field, how it has evolved, where it is going, and its commercial importance (if any).
- Brief description of the book – You should define, in a few paragraphs, what specifically the book will be about. You should discuss the approach you intend to take, for example, the balance between theory and practice, and any particular pedagogical or presentational features that will characterize the book.
- Reasons for writing the book – State the reasons you think this book should be published and how it will benefit the readers.
- Market and readership – Describe exactly who the book is aimed at, for example, graduate students, researchers, practitioners in industry, and in what subjects they work or study, for instance, electrical engineering, computer science, applied physics, mathematics. If the book can be used as a textbook then you should describe the type of course for which it could be adopted. In this section you should also describe the prerequisite knowledge that you expect of your readers.
- Competition – Give details of the main competing books (author, title, publisher) and discuss how your book will differ from them. Present the aspects that give your book an edge. This analysis is particularly important if the book you are proposing is a textbook.
- Table of contents – This should give the chapter headings along with a sentence or two explaining what each chapter will cover. You should also include the first level of subheadings. If you have a more detailed table of contents then provide it.
- Manuscript information – This should include estimates for: how long you think the book will be in printed pages, a delivery date for the finished manuscript the number of figures the book will have.
- Author, contributor information – Provide a brief resume or curriculum vitae for each author. If the book you are proposing is an

edited volume then give names and affiliations of each of the intended contributors.

- Sample writing – A detailed proposal of five pages is the minimum needed to get the assessment procedure under way. If you have any sample sections or chapters then you should send those as well.
- The assessment process – Once you have submitted the proposal it will be read by the appropriate editor. If the editor feels that the proposal is adequate then it will be sent out to a number of experts in the field. The feedback from these reviewers will be discussed with you. Their constructive feedback results in a better final book.

1.7 Writing a Book

The book is the most complete form of individual knowledge retrieval. It is the single reason mankind evolves to higher levels of technology and culture. Without books, the coded forms of writing, the human being would wake up every day to the same stage in technological and cultural evolution.

1.7.1 Initial Information

The initial information includes the title, authors, organization, editor, and date. Always capitalize the initial letters of the words of the title. The initial pages of a book also present information for librarians, copyright information, dedication, acknowledgments, information about the author, a list of figures, a list of tables, and a table of contents.

The foreword is an optional introductory statement that is written by an authority in the field. The preface announces the book's purpose, scope and context. It usually specifies the audience, contains acknowledgments of those who helped in its preparation and cite permission obtained for the use of copyrighted material. The section on organization of the text contains information on the contents of the chapters and indicates the best way to read or present them to the students.

1.7.2 Body of the Book

- The manuscript contains the chapters, which may be divided in sections and subsections. It usually includes visuals: photographs, drawings, diagrams, graphs, tables, and flowcharts, to explain the subject to the intended audience.

- Supplementary reading presents information about other books on related subjects, usually including comments from the author.
- The references form an optional list of works cited in the book.

1.7.3 Additional Information

- The appendices usually contains additional material that can be useful to the reader and clarifies or supplements the body.
- The list of abbreviations and symbols is essential in monographs, dissertations and theses.
- The bibliography is an alphabetical list of all sources that were consulted in writing the book.
- The glossary has an alphabetical list of specialized terms used in the book and their definitions.
- The index presents an alphabetical list of specialized major topics discussed in the book, along with page numbers. The index is the final section of the book.

1.8 Elements of Scientific Writing

In scientific writing some elements must be carefully considered. Most of the usual errors are related to the usage of articles, punctuation and word division.

1.8.1 Articles

Articles are used to introduce nouns, and English often requires the use of articles

- The indefinite article (*a, an*) is used when there is no need to be specific about a person or thing.
- *An* is used if the noun following begins with a vowel or sounds as a vowel (e.g., An analysis will follow. Let me show you an SMS).
- *The* is a definite article, used to identify a particular thing.

1.8.2 Punctuation

Punctuation was absent in the Antiquity, when the documents were produced in *scriptura continua*, a Latin terminology for continuous script, a style of writing without spaces, or other marks between the words or sentences. The reading of the document was aloud, and the readers used to give the proper intonation to their speech.

Punctuation, which is essential to clarify and remove any ambiguity in the meaning of sentences, is the use of spacing, conventional signs, and certain typographical devices to facilitate the understanding and the correct reading, both silently and aloud, of texts.

The first punctuation system was established for the Greeks by Aristophane of Byzantium. It used three marks or *punctus* of speech, the points, that marked the sentence parts rhetorically, by pausing a different time before beginning the next unit. The pause marks later became known as: *comma, colon* and *periodus*.

- The period (.)
  - The period is used to close sentences that make statements.
  - It is also used to form abbreviations (Ph.D., Inc., Mr.).
- The comma (,)
  - The comma is used to indicate pauses and to separate entries in lists.
  - It is also used to set off a word, phrase, or clause that is in apposition, from the Latin name *appositio*, to a noun and that is nonrestrictive.
- The semi-colon (;)
  - The semi-colon separates closely-related ideas.
  - In most cases a period should be used instead.
  - When a list includes items that have commas within them, use a semi-colon to separate the items.
- The colon (:)
  - The colon is used in headings, to announce that more is to follow.
  - It is used to introduce a list of things (words, equations).
  - It introduces a quotation.
  - When used between two clauses, it indicates that the second one provides an explanation of what was said in the first.
  - Never use the colon after the main verb in a sentence.
- The hyphen (-)
  - This mark may be used to separate two parts of a compound noun (e.g., light-year).
  - It is also used to break up words at the end of a line, and should always be placed between syllables. Proper nouns should not be broken up by hyphens.

- Parentheses ( )

  - Parentheses are used to set off an interruption in the middle of a sentence, including references to pages, diagrams, illustrations, chapters.
  - They are also used to make a point which is not part of the main flow of the sentence.
  - Use them to enclose acronyms: "Institute for Advanced Studies in Communications (Iecom)."

- The apostrophe (')

  - The apostrophe shows possession (e.g., Alencar's book on scientific style in English).
  - It shortens certain word combinations (e.g., can't, he's).
  - It must be said that contractions should be avoided in formal written work. (e.g., cannot, he is).

- Brackets [ ]

  - Brackets are used for citations or to enclose a word inserted into a quotation.

- The diagonal or dash (–)

  - A dash may be used in place of a colon, to set off a word or phrase at the end of a sentence, or an appositive to be emphasized.
  - A dash is used to summarize a thought added to the end of a sentence.

- Quotation marks (" ")

  - The main use of quotation marks is to show that the exact words written are being repeated.
  - They are also used to enclose titles of articles, chapter names, and short stories.
  - Single quotation marks (' ') can be used within a quotation.
  - A comma should precede a quotation and other punctuation should be placed inside the quotation marks.

1.8.3 Word Division

Although word division is left for the word processor nowadays, the English grammar has an adequate manner to do it. It is important to know that there are rules in English language to split words at the end of a line.

- A word should be divided only between syllables (com•pu•ter). Dictionaries indicate how to divide words at syllable breaks.
- If a vowel stands alone as a single syllable, it must remain on the same line as the first part of the word (experi•ment).
- A word is generally divided between double consonants (neces•sary), unless it means breaking up the root of the word (process•ing).
- If a word contains a natural hyphen, divide only at that point (sixty-five).
- If a word contains a prefix or suffix, it is best to divide at that point (auto•correlation).
- Do not divide one-syllable words (length), no matter how long the word may be.
- Do not divide a word in the first or last line on a page.
- Do not divide a word in the first line of a paragraph.
- Do not divide words on two consecutive lines.
- Do not divide a proper name or number (Kolmogorov, 1931).
- Do not separate two letters from the rest of the word.
- Do not separate contractions or abbreviations (wouldn't, ATSC).
- Do not use a hyphen to break a URL or an e-mail address.
- Do not use excessive word division.
- Do not separate the unit of measurement from the number (550 kHz).

1.9 Numbers, Units of Measurement and Symbols

There are strict rules for writing numbers, units of measurement and symbols. A technical manuscript is certainly refused if those rules are not followed.

1.9.1 Rules for Writing Numbers

- Write out all numbers below ten:
  - zero deviations from the expected value,
  - nine devices to count for.
- The exception to this rule are numbers used with:
  - units of measurement (3 meters),
  - age (15 years old),
  - dates (October 11, 1957),
  - time (2 seconds),
  - page numbers (page 4),

- percentages (5 percent),
- money ($8),
- proportions (30:1 or 30 to 1).

- Write the numbers as numerals if two or more are in the same section:
  - the transmitter has 5 audio amplifiers, 2 pass-band filters and a net gain of 60 dB.

- Large numbers must be written in the form most familiar to the audience:
  - 15,300,000,
  - $15,3 \times 10^6$,
  - fifteen million and three hundred thousand.

- Place a hyphen between a number and a unit of measurement when they modify a noun:
  - 3-month-old experiment.

- Use the singular for fractions and decimals that are used as adjectives:
  - 0.5 kilogram,
  - 0.1 centimeter.

- Write decimals and fractions as numerals:
  - zero point two five – 0.25.

- Do not begin a sentence with numerals.
- Keep all units consistent.

  - Chose one standard and stick with it.

- Use the correct units.
  - There are two different versions of the metric system, cgs (centimeter, gram, second) and SI (meter, kilogram, second, ampere, kelvin, mole, candela).

- When writing the units of measurement in word form they should never be capitalized:
  - hertz (Hz), ampere (A), tesla (T), ohm Ω, henry (H).

- Indicate multiplication by a raised dot ($\cdot$) and division by a slash (/):
  - 10 W/Hz,
  - 150 V$\cdot$s.

1.10 Complementary Material

It is common to misuse words and phrases in spoken English, but the scientific language is precise and the usage of words must be correct, and redundancies must be avoided.

1.10.1 Commonly Misused Words and Phrases

In the following, some commonly misused words and phrases are presented, with the correct or more adequate versions on the right hand side.

- a large number of – many
- a lot of – many
- as a general rule – like
- as shown in Table 10 – Table 10 shows
- at this point in time – at this time, now
- be considered as – as
- by means of – by
- despite the fact that – although, even though
- during the course of – during
- exhibits the ability – can
- has been widely acknowledged – is
- has proved itself to be – has proved, is
- have discussion of – discuss
- in many cases – often
- in order to – to
- in some cases, in other cases – sometimes
- in the course of –`during
- in the event that – if
- in the form of – as
- in the near future – soon, or exact date
- in the vicinity of – near
- is equipped with – has, contains
- it is clear that – clearly
- know-how – skill, ability
- kind of, sort of – rather, somehow, somewhat
- on an annual basis – yearly
- on the occasion of – when
- prior to that time – before
- start off – start

- subsequent to – after
- take action – act
- the reason why is that – because
- until such time as – until
- with reference to – about
- with the result that – so that
- deemed it necessary to – eliminate
- it has been shown that – eliminate
- it is found that – eliminate
- it is recognized that – eliminate
- it is worthy of note that – eliminate
- it may be mentioned that – eliminate
- it may be seen that – eliminate
- it must be remembered that – eliminate
- thanking you in advance for your cooperation – eliminate
- in fact that – eliminate
- what is known as – eliminate
- such (of this kind) – it

1.10.2 Redundancies that Should Be Avoided

In the midst of a conversation it is common to surrender to verbosity and duplication. In formal writing, redundancy is less excusable, but fortunately easy to correct. The following sentences should be avoided, in favor of the simpler forms.

- absolutely certain – certain
- absolutely essential – essential
- actual experience – experience
- add an additional – add
- added bonus – bonus
- adding together – adding
- advance plan – plan
- an honor and a privilege – an honor
- any and all – any
- as for example – as
- ask a question – ask
- at the present time – at present
- balance against one another – balance
- basic fundamentals – basic

- came at a time when – came when
- cancel out – cancel
- close proximity – close
- collaborate together – collaborate
- completely filled – filled
- completely finished – finished
- completely opposite – opposite
- consecutive in a row – consecutive
- continue on - continue
- current status – status
- definite decision – decision
- different varieties – varieties
- direct confrontation – confrontation
- difficult dilemma – dilemma
- during the course of – during
- end result – result
- equally as well – equally
- enter in – enter
- estimated at about – estimated
- few in number – few
- filled to capacity – filled
- final outcome – outcome
- first and foremost – first
- first introduction – introduction
- first priority – priority
- foreign imports – imports
- goals and objectives – goals
- in close proximity – close
- joined together – joined
- merge together – merge
- mixed together – mixed
- mutual cooperation – cooperation
- necessary requisite – requisite
- one and the same – the same
- overall plan – plan
- past history – history
- personal opinion – opinion
- physical size – size
- past history – history

- plan ahead – plan
- point in time – time
- postpone until later – postpone
- reason why – reason
- refer back to – refer to
- repeat again – repeat
- same identical – identical
- since the time when – since
- small in size – small
- spell out in detail – spell out
- still remains – remains
- suddenly exploded – exploded
- take action – act
- therapeutic treatment – treatment
- this particular instance – this instance
- this particular time – now
- triangular in shape – triangular
- true facts – facts
- unexpected surprise – surprise
- uniformly consistent – consistent
- unintended mistake – mistake
- usual custom – custom
- whether or not – whether
- written down – written
- you may or may not know – you may know
- where (after an equation) – in which

1.10.3 Latin Terms and Abbreviations

Terms in Latin are reminiscent of an era when the scientific literature, and every communication in the world, was written in the ancient language. The abbreviations in Latin simplify the text, and give a touch of elegance, if not abused.

- ca. – *circa* (about, approximately)
- cf. – *confer* (compare)
- e.g. – *exempli gratia* (for example)
- et al. – *et alii, et aliae, et alia* (and others)
- etc. – *et cetera* (and other things, and so on)
- et seq. – *et sequientes* (and the following)

- f.v. – *folio verso* (on the back of the page)
- ibid. – *ibidem* (in the same place)
- id. – *idem* (the same)
- i.e. – *id est* (that is)
- inf. – *infra* (below)
- loc. cit. – *loco citato* (in the place cited)
- n.b. – *nota bene* (note well)
- op. cit. – *opere citato* (in the work cited)
- Q.E.D. – *quod erat demonstrandum* (which was to be proved)
- q.v. – *quod vide* (which see, a reference to another part of a published work)
- [sic] – *sic* (so, thus, inserted in brackets)
- sup. – *supra* (above)
- s.v. – *sub verbo, sub voce* (under the word)
- ut sup. – *ut supra* (as above)
- v. *or* vs. – *versus* (inverted, against)
- viz. – *videlicet* (namely)

1.11 Chapter Summary

Scientific writing must be accurate, concise, useful, clear, illustrated with visuals, targeted to a specific audience, well-organized, interesting, consistent, complete, correct in spelling, punctuation and grammar.

A successful writing involves a few steps, which include preparation, research, organization, writing and revision. It is important to follow the plan.

This chapter introduces the definition of style, teaches the reader about scientific writing, presents guidelines for writing articles, reports and books, and how to submit a book proposal. Furthermore, it showcases the elements of scientific writing, numbers and symbols and recommends some useful complimentary material.

It is important to review punctuation, abbreviations, capitalization, contractions, dates, italics, number and measurement units, proofreading, and spelling. The following points are essential:

- Writing process – A successful writing involves several phases, which include preparation, research, organization, writing and revision.
- Research – Research involves the question: What does the researcher know about the subject?

- Organization – Organization involves the choice of the best methods to develop the work.
- Writing – During the writing phase, select an appropriate point of view, and adopt an appropriate style and tone.
- Revision – During the revision process, check for unity and coherence, verify sentence variety, emphasis, and subordination, check for ambiguity, awkwardness, and verify logic errors.

The human being evolved, culturally speaking, because of the invention of the writing. Before that, the knowledge used to be passed orally, and the technology could not be properly developed. An Egyptian scribe once wrote, circa 1300 BC,

> Be a scribe! Engrave this into your heart.
> So that your name can endure like theirs!
> The roll is better than the sculpted stone.
> A man died, his body is dust,
> and his people disappeared from the Earth.
> It is a book which makes him be remembered,
> in the mouth of the speaker who reads it.

1.12 Reference Material

Periodicals and Handbooks

- *IEEE Transactions on Professional Communication.*
    - Published since 1958 to improve professional communication among the IEEE members.
- *The Elements of Style*, by Strunk and White (Macmillan, 1979).
    - A good reference for a quick tutorial. Everybody should read this book, which presents English prose writing in general.
- *A Handbook for Scholars* by Mary-Claire van Leunen (Knopf, 1978).
    - This well-written book is a pleasure to read. It explains the use of footnotes, references and quotations.
- *Writing Mathematics Well* by Leonard Gillman. Mathematical Association of America.
    - Gillman's book refers to the three previous classics in the field:

* An article by Harley Flanders, *Amer. Math. Monthly*, 1971, pp. 1–10.
* Another by R. P. Boas in the same journal, 1981, pp. 727–731.
* There is also a nice booklet called *How to Write Mathematics*, published by the American Mathematical Society in 1973, especially the essay by Paul R. Halmos on pp. 19–48.

- *Mathematical Writing*, by Donald E. Knuth, Tracy Larrabee, Paul M. Roberts. Mathematical Association of America, 1989.

  – This book is based on a course given by Donald Knuth at Stanford University. It discusses technical writing and presentation of mathematics and computer science, including preparation of theses, papers, and books.

Reference Books

- Handbook of Technical Writing, Gerald J. Alred, Charles T. Brusaw, Walter. E. Oliu. St. Martin's Press, New York, 2006.

  – Recommended; complete handbook on technical writing; entries arranged in alphabetical order; excellent index; English as a second language guidance; includes succinct guide to the writing process. Material found at `bcs.bedfordstmartins.com/alredtech`

- An Outline of Scientific Writing: For Researchers With English as a Foreign Language, J. T. Yang. World Scientific, 1995.

  – May be especially useful to English as a second language writers

- MIT Guide to Science and Engineering Communication, J. G. Paradis and M. L. Zimmerman. MIT, Cambridge, 2002.

  – Discusses all types of technical communication and includes a list of 27 guidelines for style and usage.

- How to Read a Book, Mortimer Adler and Charles van Doren. MJF Books, New York, USA, 1972. ISBN 1-56731-010-9.

  – A classic book describing the four stages of reading, includes a list of relevant books.

- *English for Electrical Engineers*, by James MacAllister and Giorgio Madama. Longman, 1976.

  – The book presents and analyzes interesting texts, in increasing order of difficulty, along with language tips and grammar.

Popular Words and Spelling

- *The Longman Dictionary of Contemporary English.*

  - Instead of the historical words found in the previously mentioned dictionaries, this one has the popular words explained in very simple English.

- *Webster's New Word Speller Divider.*

  - People who do not spell well find this book to be quite useful.

- *Roget's Thesaurus.*

  - This book is a synonym dictionary. Use it if you know that a word exists but you have forgotten it or when you want to avoid repetition, to define a new technical term or a new name for a paper.

- *The Secretary's Quick Reference Handbook*, by Sheryl L. Lindsell. Arco, 1989, New York.

  - This is a simplified and easy-to-use manual to correct writing.

The TEX and LATEX Reference Books

- The TEXbook (Computers & Typesetting), by Donald E. Knuth. Addison-Wesley Professional. 1984, New York, United States.

  - This is the reference book on TEX

- LATEX A Document Preparation System (2nd Edition), by Leslie Lamport Addison-Wesley Series on Tools and Techniques for Computer T, 1994, New York, United States.

  - This is the reference book on LATEX

Dictionaries

- *Webster's Dictionary of English Usage*, 1989.

  - A dictionary is always a wonderful resource. This edition goes well beyond the *American Heritage* usage notes. It is filled with choice examples.

- *Webster's Standard American Style Manual.* Merrian-Webster Inc. 1985, Springfield, United States.

- A practical guide to the conventions of the English language, which offers concise and comprehensive descriptions of the rules to prepare documentation that is clear and consistent.

- *The Oxford English Dictionary* (usually called the OED).

  - This is a useful reference book.

- The *OED Supplement.*

  - The supplement brings the OED up to date. The supplement comes in four volumes.

- *The American Heritage Dictionary*

  - This dictionary contains usage notes and an appendix with Indo-European root words.

- *Merriam-Webster Dictionary and Thesaurus,* can be found at `www.m-w.com`.

  - Useful, presents etymology and pronunciation of words.

- *Wordsmyth Dictionary and Thesaurus,* can be found at `www.wordsmyth.net`.

  - Very useful, although definitions are brief, it identifies parts of speech.

Dictionaries and Grammars

- *The Encyclopedic Dictionary of Style and Usage,* by Mary A. DeVries. Berkley Books, 1999, New York, United States.

  - A compact dictionary, with cross-references in place of a traditional subject index.

- *The Oxford English Dictionary of English Grammar,* by Sylvia Chalker and Edmund Weiner. Oxford University Press. 1994, London, United Kingdom.

  - This is a very useful reference book.

- *Modern Grammar,* by Paul Roberts. Harcourt, Brace & World, Inc. 1968, New York, United States

  - This book presents the sentence-producing mechanism and examines the skeleton of the English sentence.

- *Latin Grammar*, by Frederic M. Wheelock. HarperCollins Publishers, Inc. 1992. New York, USA.
  - This book presents the basics of Latin, in case the reader is interested in knowing the roots of the words.

Recommended Guides to Writing

- *The Elements of Technical Writing*. Gary Blake and Robert W. Bly. Macmillan. 1993. United States.
  - A concise reference text with useful information on clear writing and punctuation.
- *The Canadian Style – A Guide to Writing and Editing*, by Dundurn Press Limited in cooperation with Public Works and Government Services Canada. 1997, Toronto, Canada.
  - A reference text with useful information for public servants on standards, recommendations and information to ensure quality in their writing.
- *The Broadview Pocket Guide to Writing*. Doug Babington and Don LePan. Broadview Press, 2002. Canada.
  - This book addresses the specialized writing problems of professionals, including information on units of measurement, equations and symbols, principles of technical communication, and how to write proposals, articles, reports and manuals.
- *How to Write and Publish a Scientific Paper*. Robert A. Day. Cambridge University Press, 1979. Cambridge, UK.
- *Alphabet to email: How Written English Evolved and Where It's Heading*, by Naomi S. Baron. Routledge, London, United Kingdom, 2000.
- *Pause and Effect*, M. B. Parkes. Scholar's Press. Aldershot, England, 1992.

Recommended Online Writing Guides

- *Mayfield Handbook of Technical and Scientific Writing*. L. C. Perelman, J. Paradis, and E. Barrett; `mit.imoat.net/handbook/home.htm`.

- Recommended as a complete guide to technical writing from MIT; concise explanation of most aspects of technical writing. It gives English as a Second Language (ESL) pointers.

- *Online Writing Lab (OWL).* owl.english.purdue.edu/handouts.

  - Guide to effective writing at college level. Includes grammar and punctuation with exercises and ESL.

- Supporting material for *Handbook of Technical Writing.* Alred et al.; bcs.bedfordstmartins.com/ alredtech.

- *Grammar, Punctuation, and Capitalization: A Handbook for Technical Writers and Editors.* stipo.larc.nasa.gov/sp7084.

  - NASA Report. Hypertext or PDF. Includes rules for technical writing.

- *AIP Style Manual.* public.lanl.gov/kmh/AIP_Style_4thed.html.

  - American Institute of Physics gives stylistic guidance, especially relevant to physics articles.

- LaTeX Style Guide for the *Journal of Integer Sequences* – Version 1.10.
- Wikipedia, the free encyclopedia. Web-page: en.wikipedia.org.

Writing References

- Bartleby Classic Online Books www.bartleby.com.
- A good collection of writers' aids, that includes the American Heritage Dictionary, Roget's Thesaurus, quotations, and more:

  - *American Heritage Dictionary.* www.bartleby.com/61.
  - *Elements of Style.* www.bartleby.com/141, classic handbook, written by William Strunk in 1918.
  - *King's English.* www.bartleby.com/116, by H. W. Fowler (1908), another classic.

- *Writing Research Papers – A Complete Guide,* by James D. Lester. Scott, Foresman and Company, 1984, Glenview, United States.

  - A guide for college and university students inexperienced in research paper writing.

- *How to Edit a Scientific Journal,* by Claude T. Bishop. ISI Press, 1984, Philadelphia, United States.

Formal Report References

- The Random House Guide to Business Writing, by Forman, Janis and Kelly, Kathleen (1990). McGraw-Hill. New York, USA.
- Reporting Technical Information, by Kenneth W. Houp and Thomas E. Pearsall (1992). MacMillan Publishing Company, New York, USA.
- Advanced Business Communication, by John M. Penrose, Robert W. Rasberry and Robert J. Myers (1993). Wadsworth Publishing Company. Balmont, California, USA.

Recommended Guides for Theses and Dissertations

- How to Write a Good Ph.D. Thesis and Survive the Viva, by Stefan Rüger. Knowledge Media Institute, The Open University, Buckingham, United Kingdom, 2016.
- Doing postgraduate Research in Australia, by K. Stevens and C. Asmar. Melbourne University Press, Melbourne, Australia, 1999.
- How to Get a PhD: A Handbook for Students and their Supervisors, by E. M. Phillips and D. S. Pugh. Open University Press, Buckingham, England, 1994.
- The Visual Display of Quantitative Information, by E. R. Tufte. Graphics Press LLC, Cheshire, United States, 1983.
- Envisioning Information, by E. R. Tufte. Graphics Press LLC, Cheshire, United States, 1990.

2

Mathematical Style in English

"Pure mathematics is, in its way, the poetry of logical ideas."
Albert Einstein

2.1 Introduction

Mathematics is an important and unique characteristic of human thought. Of all the sciences, only in Mathematics there is no significant correction, but only extension, over time. Once the deductive method was developed by the Greeks, their results remained correct for all time.

For example, no one disputes Pythagoras' theorem, a fundamental relation in Euclidean geometry among the three sides of a right triangle, a mathematical result for which the proof was developed more than 2500 years ago.

The mathematical proof remains true, although some historians have found evidence that the Pythagorean theorem was known to the mathematicians of the First Babylonian Dynasty (20th to 16th centuries BC), which would have been over a thousand years before Pythagoras (ca. 570–495 BC) was born.

The origins of Mathematics derive from the concepts of number, magnitude, and form, but the discipline may also be defined as the study of relationships among quantities, magnitudes and properties, and of the logical operations by which they are related and may be deduced.

The main characteristic of mathematical writing is its conciseness, because the mathematical notation was developed to save time and space. A long and verbose explanation can typically be replaced by a nice and compact formula.

The purpose of this chapter is to help with writing scientific texts in English, more specifically when using mathematical terms, including style recommendations to enhance readability, precision and conciseness.

The reader will find helpful tips to avoid commonly made mistakes, related to mathematical notation, important rules to remember when writing, and specific guidelines for Linux and Latex users.

2.2 Definition

The mathematical style has the objective of enabling the readers to understand mathematical concepts, use the right notation, and write properly.

As the reader becomes a writer, the author usually develops a particular manner or technique by which he creates the texts and expresses himself. Anyway, because the subject is more important than the writer's opinion, the mathematical style generally uses an objective tone.

2.3 Mathematical Style in English

2.3.1 Mathematical Communication

The following sections cover the basics when writing technical or scientific works, which include any type of numbers.

Four aspects of good mathematical communication are:

- Semantics – is a linguistic and philosophical study of meaning, either in language, programming languages, formal logic, and semiotics. It focuses on the relationship between signifiers, such as, words, phrases, signs and symbols, and what they indicate. It is related to words, and the job they do in the sentences.
- Syntax – is the set of rules, principles, and processes that govern the structure of sentences in a given language, specifically word order and punctuation. Also known as grammar.
- Symbol – is a mark, a sign, or a word that represents an idea, an object, or a relationship between objects in a mathematical sentence. Symbols simplify the language and creates connections between different concepts. They are very meaningful to mathematical writers.
- Style – is a manner of doing or presenting concepts. It is, in a sense, a synthesis of the above.

2.3.2 Mathematical Reading

It is assumed that the first mathematical symbols were signs to depict numbers, the ciphers, which probably preceded the creation of written language.

The Babylonian and the Egyptian were the most ancient systems of numbering, dating back to 3500 B.C..

Symbols

The creation of the modern algebraic symbols dates back to the 14th century, at the time related to practical arithmetic and the study of equations, but the first mathematical symbols for arbitrary quantities appeared only in the 5th century B.C., in Greece.

The French mathematician René Descartes (1637) improved the algebraic notation, denoting unknowns by the last letters of the alphabet x, y, z, and arbitrary given quantities by the first letters a, b, c. Descartes is also credited with the modern notation for powers, and with the popularization of the Cartesian coordinate system.

Cardinal numbers, or cardinals, are a generalization of the natural numbers used to measure the cardinality, or size, of sets. When reading cardinal numbers in English, keep the following examples in mind:

Numerals

A numeral system is a writing system for expressing numbers. It is a mathematical notation for representing numbers in a set, using digits or other symbols.

How to read Cardinal numbers:

- 103 – a hundred and three.
- 200 – two hundred.
- 1,657 – a thousand six hundred and fifty-seven.

Ordinals

An ordinal number, or ordinal, is a generalization of the concept of a natural number that is used to describe a way to arrange a collection of objects in order, one after another. Ordinal numbers are the "labels" needed to arrange collections of objects in order.

How to read ordinal numbers? When reading ordinal numbers in English, keep the following examples in mind:

- 103rd – a hundred and third.
- 200th – two hundredth.
- 1,657 – one thousand six hundred and fifty-seventh.

Decimals

This section explains how to read the decimals in English. Some examples are presented in the following:

- 20.82 – twenty point eight two (centesimals).
- 1.444... or $1.\overline{4}$ – one point four recurring.
- 1.06 – one point zero, or "o", six. Also, one point six decimals.
- 0.37 – naught point three seven (centesimals).
- 0.002 – naught point zero zero two, or zero point two millesimal.

Fractions

The term fraction came from the Latin word for broken, *fractus*, and represents a part of a whole or any number of equal parts. This section explains how to read fractions in English. Some examples follow:

- $\frac{2}{3}$ – two over three.
- $1\frac{1}{2}$ – one and a half.
- $1\frac{1}{3}$ – one and one third.
- $\frac{2}{4}$ – two quarters, two fourths.
- $\frac{1}{10}$ – a or one tenth.
- $\frac{1}{48}$ – one over forty-eight.
- $\frac{1}{k}$ – one over k, the reciprocal of k.

Operation

An operation is a calculation of input values, called operands, to obtain an output value. This section explains the reading of the basic mathematical operations, with some examples.

- $5 + 4 = 9$ – five plus four equals nine.
- $5 - 4 = 1$ – five minus four equals one.
- $5 \times 4 = 20$ – five times four equals twenty.
- $5/4 = 1.25$ – five divided by four equals one point two five.
- 5^2 – five squared.
- $\sqrt{81} = 9$ – the square root of eighty-one is nine.
- $5^3 = 125$ – five cubed equals one hundred and twenty-five.
- $\sqrt[3]{64} = 4$ – the cubed root of sixty-four is four.
- 5^6 – five to the power six.
- 10^{-2} – ten to the minus two.

- (a, b) – the open interval $a < x < b$.
- $(a, b]$ – the half-open interval $a < x \leq b$.

Algebraic Operations

An algebraic operation involves the traditional computations of arithmetic, which are addition, subtraction, multiplication, division, raising to an integer power, and taking roots. The following examples show how to write and read algebraic operations:

- All variables must be defined when they are first introduced.
- $j = \sqrt{-1}$ – the positive square root of minus one.
- x_k – x sub k (subscript).
- y^n – y to n (superscript).
- $f(t)$ – f of t, f is a function of t.
- f^{-1} – inverse of the function f.
- f' – derivative of f.
- f'' – second derivative of f.
- $\frac{df}{dx}$ – derivative of f with respect to x.
- $\frac{\partial f}{\partial x}$ – partial derivative of f with respect to x.
- $|y|$ – absolute value of y.
- $\lceil y \rceil$ – greatest integer not greater than y.

Equation, Formula, Expression, Series

Not all mathematical objects are equations. It is important to acknowledge, in scientific texts, the difference between an equation, a formula, an expression, and a series. This section contains examples of equations, formulas, expressions and series, to help the readers.

- $ax^2 + bx + c = 0$ – Equation.
- $x_i = \frac{b \pm \sqrt{b^2 - 4ac}}{2a}$, i=1,2 – Formula.
- $y = ax^2 + bx + c$ – Expression.
- $y = a_1 x^1 + a_2 x^2 + a_3 x^3 + \cdots$ – Series.

Percentages

A percentage is a number, or ratio, expressed as a fraction of one hundred. It is a dimensionless, or pure, number. This is how to read and write percentages:

- 2.5% – two point five per cent.
- 6% – six per cent.

- 100% – one hundred per cent.
- 340% – three hundred and forty per cent.

Calendar Dates

A calendar date is a reference to a particular day, month or year, represented within a calendar system. For instance, the calendar date permits a certain day to be identified. How to read and write dates:

- 1683 – sixteen eighty-three.
- 1800 – eighteen hundred.
- 1801 – eighteen hundred and one.
- 1810 – eighteen ten.
- 1956 – nineteen fifty-six.
- 2009 – two thousand and nine.
- 2056 – twenty fifty-six.
- 2156 – twenty-one fifty-six.

A comma is not used to separate the digits in a date. The dates should be expressed in the same terms. If someone needs to say "1637", then say "late in the 1500s", not "late in the 16th century".

It is important to check the American versus the British usage, as in the following examples:

- June 1st, 2009 – June first, two thousand and nine (US).
- 1st June 2009 – the first of June, two thousand and nine (United Kingdom).
- June 25, 1997 – June twenty-fifth, nineteen ninety-seven (US).
- 25 June 1997 – the twenty-fifth of June, nineteen ninety-seven (United Kingdom, US Military).

The Greek Alphabet

The Greek alphabet has been in use for more than eighteen centuries, mainly to write the Greek language. It was derived from the earlier Phoenician alphabet, and was the first alphabet to have distinct letters for vocalic and consonantal sounds. In the past, only the consonantal sounds used to be represented. It is the ancestor of the Latin and Cyrillic scripts.

The Greek alphabet is a source of technical symbols and labels in many fields of mathematics and science. This section shows how to read and write the Greek letters, both in capital or lowercase.

Name	Capital	Lowercase
alpha	A	α
beta	B	β
gamma	Γ	γ
delta	D	δ
epsilon	E	ϵ
zeta	Z	ζ
eta	H	η
theta	Θ	θ, ϑ
iota	I	ι
kappa	K	κ
lambda	Λ	λ
mu	M	μ
nu	Nu	ν
xi	Ξ	ξ
omicron	O	o
pi	Π	π
rho	R	ρ
sigma	Σ	σ, ς
tau	T	τ
upsilon	Υ	υ
phi	Φ	ϕ, φ
chi	X	χ
psi	Ψ	ψ
omega	Ω	ω

2.4 Mathematical Writing

2.4.1 Formulas and Theorems

This section provides some insight into writing formulas and theorems, with examples of good and bad practices.

First, consider that the text improves if symbols in different formulas are separated by words.

- Incorrect: Consider $\frac{p}{q}$, $q > 0$.
- Correct:: Consider $\frac{p}{q}$, in which $q > 0$.

Do not begin a sentence with a symbol.

- Incorrect: $x^n - a = 0$ has n distinct zeroes.
- Correct:: The equation $x^n - a = 0$ has n distinct zeroes.

Use the right words.

- Do not over-use the symbol ($\therefore$).

- Replace it by "therefore", except in texts on set theory.

The statement just preceding a theorem or algorithm should be a complete sentence or should end with a colon.

- Incorrect: One now has the following

    - **Theorem**. $F(t)$ is continuous.

- Correct: One can now prove the following result:

    - **Theorem**. The function $F(t)$ defined in (10) is continuous.

It is better to replace the first sentence by a more suggestive one, relating the theorem with the previous discussion.

Theorem

A theorem is a statement that has been proved based on established statements, such as other theorems, and generally accepted, or fundamental statements, called axioms. The statement of a theorem must be self-contained, and does not depend on the assumptions of the preceding text.

In the following section there are some tips on writing mathematical sentences:

2.4.2 Sentences

Be careful with the prose when writing the sentences. Read what has been written, and change the wording if it does not flow smoothly.

Do not omit "that" when it helps the reader to parse the sentence.

- Incorrect: Assume A is a set.
- Correct: Assume that A is a set.

The words "assume" and "suppose" should usually be followed by "that", unless another "that" appears in the sentence. On the other hand, to not overuse "that."

- Incorrect: We have that $x = y$.
- Correct: We have $x = y$.

Also avoid the expression "because of the fact that." It is better to write "based on the fact."

Vary the sentence structure and the choice of words, resorting to synonyms, when necessary, to avoid monotony. It is important to use parallelism when parallel concepts are being discussed.

- Incorrect: Formerly, science was taught by the textbook method, while now the laboratory method is employed.
- Correct: Formerly, science was taught by the textbook method; now it is taught by the laboratory method.

The word "we" is useful to avoid the passive voice. But "we" should be used in contexts in which it means "you and me together", *not* a formal equivalent of "I." In most technical writing the first person should be avoided, unless the author's persona is relevant.

2.4.3 Words to Avoid

Avoid words, such as, "this" or "also" in consecutive sentences; such words, as well as unusual or polysyllabic utterances, tend to stick in a reader's mind longer than other words, and good style will keep "sticky" words spaced well apart.

Tie the concepts and formulas together with a running commentary. Do not use the style of homework papers, in which a sequence of formulas is merely listed, with no comments. Try to state things twice, in different or complementary ways, especially when giving a definition, to reinforce the reader's understanding.

2.4.4 Motivation Is Important

The most important principle of good writing is to keep the reader as the target of one's writing. What does the reader know so far? What does the reader expect next and why?

When describing the work of other people, it is sometimes safe to provide motivation by simply stating that it is "interesting" or "remarkable".

But it is best to let the results speak for themselves or to give *reasons* why the things seem interesting or remarkable. When describing one's own work, be accurate and do not use superlatives of praise, either explicitly or implicitly, albeit being enthusiastic.

2.4.5 Readers Glance over Formulas

Many readers will skim over formulas at first, therefore, the sentences should flow smoothly when all, but the simplest formulas, are skipped.

Do not use the same notation for two different things, conversely, use consistent notation for the same thing when it appears in several places.

- For example, do not say "x_j for x_n" in one place and "x_k for x_n" in another place unless there is a good reason

It is often useful to choose interesting and meaningful names for indices, such as, k varies from 1 to K, and to maintain a consistent usage. Typographic conventions, such as lowercase letters for elements of sets, uppercase for sets and calligraphic letters for families of sets, are also useful.

2.4.6 Subscripts

Do not get carried away by subscripts, especially when dealing with a set that does not need to be indexed; set element notation can be used to avoid subscripted subscripts.

For example, it is often troublesome to start out with a definition such as "Let $X = \{x_1, \ldots, x_n\}$" if there is going to be a need for subsets of X, since the subset will have to be defined, for example, as $\{x_{i_1}, \ldots, x_{i_m}\}$.

Also there may be the need to speak of elements x_i and x_j all the time. Do not name the elements of X unless necessary. Then refer to elements x and y of X in the subsequent discussion, without needing subscripts. Or refer to x_1 and x_2 as specified elements of X.

2.4.7 Formulas and Sentences

Show important formulas on a line by themselves. If there is a need to refer to these formulas from remote parts of the text, reference numbers are given to the most important ones, even if they are not referenced. Sentences should be readable without ambiguity.

Incorrect examples: "Alencar remarked in an article about the importance of probability." "In the theory of sets, fields and other algebraic structures are analyzed."

Numbers below ten should be spelled out when used as adjectives, but not when used as names.

- Incorrect: The method requires 3 passes.
- Correct: Method 5 is illustrated in Figure 3. It requires three passes.

Capitalize the important mathematical objects: Theorem 1, Lemma 2, Algorithm 3, Method 4, Equation 5, Table 6, Figure 7.

2.5 How to Deal with Numbers

It is not always easy to deal with numbers, because there is a number of ways to express them.

Number Versus Numeral

What is the difference between a number and a numeral? A number is an abstract concept while a numeral is a symbol used to express that number. "Six," "6" and "VI" are all symbols used to express the same number (or the concept of "sixness"). It is possible to relate the difference between a number and its numerals to the difference between a person and her name.

Spell Small Numbers Out

The whole numbers smaller than ten should be spelled out. That is a definite rule for a more formal approach in writing.

Experts do not always agree on all rules. Some experts say that any one-word number should be written out. Two-word numbers should be expressed in figures. That is, they say one should write out twelve or twenty. But not 24. But it is safer to express numbers above ten using numerical digits.

Usage of the Comma in Technical Writing

In English, the period is the decimal separator. The comma is used as a thousands separator, to make large numbers easier to read. Therefore, write the size of Brazil as 8,515,767.049 km², instead of 8515767.049 square kilometers.

In Continental Europe and Latin America, the opposite is true, periods are used to separate large numbers and the comma is used for decimals. Finally, the International Systems of Units (SI) recommends that a space should be used to separate groups of three digits, and both the comma and the period should be used only to denote decimals, like $10 500,40.

How to Start a Sentence

Do not start a sentence with a numeral. Write "Five score and six years ago," not "5 score and 6 years ago." That means one might have to rewrite some sentences: "Readers bought 5,000 copies the opening day" instead of "5,000 copies were sold the opening day."

Ages, Percentages and Other Numbers

Centuries and decades should be spelled out. Use "the Eighties" or "nineteenth century."

Sometimes it is necessary to to deal with percentages and recipes. With everyday writing and recipes it is possible to use digits, such as, "They are 6% of the players" or "Add 2 cups of sugar." In formal writing, however, the author should spell the percentage out like "They are six percent of the players".

If the number is rounded or estimated, it should be spelled out. Rounded numbers over a million are written as a numeral plus a word. Use "About 450 million people speak Spanish natively," instead of "About 450,000,000.00 people speak Spanish natively." If the author is using the exact number, it should be written out.

If two numbers are next to each other, it can be confusing if the author writes "5 16-year-olds", therefore, write one of them as a numeral, such as, "five 16-year-olds."

With ordinal numbers it is important to to maintain consistency. An author should not write "This was my 1st book," but rather "This was my first book." It is necessary to be consistent within the same sentence. If an industry has 15 interns, it must have 15 supervisors, not fifteen supervisors.

2.6 Miscellanea

2.6.1 How to Deal with Words

Some words that are frequently misspelled, and the author must take care of the quality of the text. This section contains some examples of commonly misspelled words:

- Correct: implement – Incorrect: impliment.
- Correct: complement – Incorrect: compliment.
- Correct: occurrence – Incorrect: occurence.
- Correct: dependent – Incorrect: dependant.
- Correct: auxiliary – Incorrect: auxillary.
- Correct: feasible – Incorrect: feasable.
- Correct: preceding – Incorrect: preceeding.
- Correct: referring – Incorrect: refering.
- Correct: category – Incorrect: catagory.
- Correct: consistent – Incorrect: consistant.

- Correct: descendant (noun) – Incorrect: descendent.
- Correct: its (belonging to it) – Incorrect: it's (it is).

2.6.2 Usage of Words

The following words are no longer being hyphenated in current literature:

- nonnegative, nonzero

Do not say "which" when "that" sounds better. The general rule is to use "which" only when it is preceded by a comma or by a preposition, when the clause is nonrestrictive, or when it is used interrogatively.

- Incorrect: Do not use commas which are not necessary.
- Correct: Do not use commas that are not necessary.

Do not use "where" when the meaning is "in which". For example, after an equation to describe the parameters.

- Incorrect: where ω_c is the carrier frequency.
- Correct: in which ω_c is the carrier frequency.

Another common error is to say "less" when the proper word is "fewer".

2.6.3 Use of Italics

It is a good practice to use italics in scientific texts.

- For foreign words.
- For variables in equations, formulas and expressions.
- For the names of variables when they appear in comments.
- Functions must be written in roman. For example: cos, exp.

2.6.4 Common Mistakes

A very common error is the misplaced "only." To illustrate, take the sentence "I had a meeting with my supervisor this morning." and insert "only" in each of the ten possible positions. Each resulting sentence carries a different meaning:

- Only I had a meeting with my supervisor this morning.
- I only had a meeting with my supervisor this morning.
- I had only a meeting with my supervisor this morning.
- I had an only meeting with my supervisor this morning.
- I had a meeting only with my supervisor this morning.

- I had a meeting with only my supervisor this morning.
- I had a meeting with my only supervisor this morning.
- I had a meeting with my supervisor only this morning.
- I had a meeting with my supervisor this only morning.
- I had a meeting with my supervisor this morning only.

2.7 Important Parts

The opening paragraph should be the best paragraph, and its first sentence should be the author's best sentence. It can make a difference, and may define if the reader will read the rest of the chapter.

If the introduction is bad, the reader will be resigned to fight with the prose. Conversely, if the beginning flows smoothly, the reader is attracted to the text and will forgive occasional lapses in the other parts. The worst way to start is with a sentence of the form:

- An x is defined as y.

Another example, with the incorrect and the correct styles.

- Incorrect: An important method for estimation uses the linear mean square estimator.
- Correct: Linear mean square is an important method for estimation, because . . ..

2.7.1 Important Rules

Commas and Periods

The style rules for English state that commas and periods should be placed inside quotation marks, but other punctuation (such as colons, semicolons, question marks, exclamation marks) stay outside, unless they are part of the quotation, except when using quotation marks to describe some text as a specific string of symbols. For example:

- Incorrect:: A sentence like "A comma is needed", is an example.
- Correct:: A sentence like "A comma is needed," is an example.

Punctuation is logical with respect to parentheses and brackets. A period should be placed inside parentheses if and only if the sentence ending with that period is entirely within the parentheses. For example:

- Incorrect: This an incorrect example (as can be noted.)
- Incorrect: This a correct example (as can be noted).

Temptations and Common Mistakes

It is necessary to resist the temptation to use long strings of nouns as adjectives: "Consider the packet switched data network protocol problem." In general, do not use jargon unnecessarily. Even specialists in a field get more pleasure from papers that use a common vocabulary.

2.8 Comments and Tips

It is important to realize that not all formulas are equations.

- Depending on the formula, the terms "expression", "relation", "definition", "series", "statement", or "theorem" might be used.
- Be careful to distinguish between mathematical notation and programming language notation.
    - While it may be appropriate to use $f[r]$ in a program, in a formal paper it is better to use f with a subscript of r.
- As another example, it is incorrect to use a star, or asterisk, $(*)$ to denote multiplication.
    - One should write the multiplication of a times b, as ab or $a \cdot b$, or $a \times b$.
- It is usual to call a an element of A and a_k an element of A.
- It is better to call a_k a "component" of A, thus distinguishing two types of subsidiary relationships.

Remember to put words between adjacent formulas, to permit the reader to follow the prose and understand the deduction.

- Ellipses, such as $(x_1, \ldots, x_n)$, require commas before and after the three dots.
- When placing ellipses between commas, the three dots belong on the same level as the commas. However, when the ellipsis is bracketed by symbols such as '$+$' or '$<$', the dots are placed at mid-level.
- Be careful with the spacing around ellipses. The character string "..." looks strange (it should have more space after the last dot).
- All kinds of accidents happen concerning spaces in formulas.
- Typesetting systems, such as, LATEX have built-in rules to cover most of the cases, but if the text is basically mathematical, strange things can appear.

2.8.1 Line Breaks and Formulas

Line breaks in the middle of formulas are undesirable, and also look bad. There are ways to enforce this using the a LaTeX command.

People who use LaTeX and wish to use the vertical bar and the empty set symbol in notation such as "$\{a \mid a \in \emptyset\}$" look for the LaTeX commands \mid and \emptyset.

2.8.2 Numbering Formulas

Numbering all displayed formulas is not recommended. Number the important ones only, those that should be referred to. Extra parentheses can also be distracting.

- In the phrase "let x be $(y + b)$," the parentheses should be omitted.

Typographic tools can be overused by having too many different fonts in one paper. It is important to pay attention to the use of colons. While the colon in:

- "Define it as follows:" is correct.

The one in:

- "One has: $s(t) = a(t)\cos(\omega_c t + \phi)$," should be omitted since the formula just completes the sentence.

Do not use a colon before a single equation, expression, or formula. Use a colon before a list or an explanation that is preceded by a clause that can stand by itself.

2.8.3 Colon and Commas

Note how to use colons and commas correctly.

- Should the first word after a colon be capitalized?
  - Yes, if the phrase following the colon is a full sentence.
  - No, if it is a sentence fragment.
- An excess of commas interferes with the smooth flow of a sentence, but too few can make a sentence difficult to read.

2.8.4 Parentheses

There are a few rules to keep in mind when using parentheses:

- Putting too many objects in parentheses is a stylistic option that can be confusing.

- Among the parentheses that can be removed, are nested parentheses. Therefore, it is better to write "(Definition 2)" than "(Definition (2))".
- In some cases the reader may expect nested parentheses. In this case think about changing the outer pair to brackets or curly-braces, in this order.
- This was the convention, in the past, but it is now obsolete, because brackets and curly braces have semantic content for some scientific disciplines.

2.8.5 Proofs

Come guidelines when writing mathematical proofs:

- An entire paper or proof in capital letters is distracting. It gives the impression of sustained shouting, as used in social networks. The same observation is valid for boldface, italics etc.
- Paul Halmos introduced the convention of placing a box at the end of a proof; this box serves the same function as the initials for the Latin expression *Quod Erat Demonstrandum* (Q.E.D.), which means "what was to be demonstrated". Leave a space between the box and the final period.
- It is important to make it clear that a new paragraph has begun. When using displayed formulas, this can become confusing unless the author is careful.

2.8.6 Line Breaks

Avoid line breaks, mainly in books, as much as possible.

- Do not let the final symbol lie on a line separate from the rest of its sentence.
- Never separate a unit of measure from its quantity.
- Do not let a reference be on a line by itself.

When using LaTeX a tilde (~) in place of a space will cause the two symbols on either side of the tilde to be tied together. It is also useful for citations.

2.8.7 Words and Symbols

Good practices to keep in mind when writing words together with symbols:

- Minimize the use of subscripts. For example, "x_i is an element of X" could more easily be "x is an element of X".

- Remember to place words between adjacent formulas.
- Be careful to define symbols before using them.
- Do not rely just on one or two styles of sentences. The following introductory words can become very monotonous:
  - Thus, ...
  - Consequently, ...
  - Therefore, ...
  - And so, ...

Is it important to be precise. One could use imprecise wording in oral communication, but written communication must be precise.

- Be precise in the wording. To mean "not decreasing," do not write "increasing".
- The former means that $x_j \geq x_{j+1}$ for *some* j, while the latter that $x_j > x_{j+1}$ for *all* j.
- Avoid the use of mixed tenses on the same subject. For example, "We assumed this to show a contradiction," is better than, "We assume this and show that it leads to a contradiction".
- In place of "Assume by contradiction that ...", it is better to say "The proof that ... is by contradiction," and even better to say "To prove ..., assume the opposite and ...".

2.9 Preparing Books for Publication

2.9.1 Revision of the Text

Search for the right and precise words to use in the text.

- When using the word "instead", be clear about the contrast that is implicit.
- The reader should immediately understand what it is being referred to:
  - Correct: And if $x = -1$ instead, ...
  - Incorrect: And if $x = -1$ instead of $+1$, ... Notice the helpful use of a redundant '+' sign.
- Avoid repeating words in a sentence or paragraph. Use synonyms or express the idea in a different way.

Look for the right tense to use in the sentences.

- Use the present tense for timeless facts. Things that were proved some time ago are nevertheless still true.

- Breaking existing paragraphs into smaller paragraphs can also be helpful.
- While editing for flow, sentences can be broken up by changing semicolons to periods.
- Make sure the variable names are not misleading. Variable names that are too similar to conceptually unrelated variables can be confusing.
- Systematic variable renaming is one of the advantages of text editors.
- Sometimes moving a formula from the embedded text to one separately displayed allows the formula to be more logically divided.

2.9.2 Linux Users

- Linux users can profit from the command `ispell` to revise the Manuscript:

  ```
  - ispell manuscript.tex
  ```

- The command identifies strange words and prompts a change or replacement, using entries from an internal dictionary.

2.9.3 Fractions

These are some of the most common errors when writing mathematical works containing fractions:

- One of the common errors that mathematicians make when they typeset is to over-use the form of fraction with a horizontal bar $\frac{1+x}{y}$ rather than a slash $(1+x)/y$.
- The stacked form can lead to tiny little numbers, especially when they are used in exponents.
- One of the common changes that mathematical copy editors make is to slash fractions.

2.9.4 Exercises

Exercises are some of the most difficult parts of a book to write. In the following there are some facts to keep in mind when writing exercises.

- Since an exercise has very little context, ambiguity can be a problem. A little redundancy can be helpful for the reader.
- For this reason, exercises are also the hardest technical writing to translate to other languages.

- Copyright law has changed, making it necessary to give credit to all previously published exercises.
- Tracing the history of even well-known theorems can be difficult, because mathematicians usually omitted citations.
- We can dispense with some of our rhetorical guidelines when writing the answers to exercises. Answers that are quick and that begin with a symbol are acceptable.

2.10 The LaTeX Style

2.10.1 How to Prepare Papers in LaTeX

Authors that prepare papers in LaTeX should observe the following guidelines:

- Avoid the use of special-purpose macros whenever possible.
- Remove all references to any packages that are not actually used. (Do *not* just comment them out.)
- It may be worthwhile to download the LaTeX file for a paper already published in the journal and model the paper on it.
- Do not include a date in the paper.
- Acknowledgments should be in a separate, numbered section at the end of the paper.

2.10.2 Common Grammatical Errors

- Avoid the passive voice, if possible.
- Instead of saying "In [9] it is shown that all primes > 2 are odd", say "Madeiro showed that all primes > 2 are odd [9]".
- Avoid the use of vague expressions such as "this number". For example:
  - Incorrect: Let x be a prime. We now square this number.
  - Correct: Let x be a prime. We now square x.
- Avoid the use of contractions: "don't", "can't", "isn't", etc.
  - Incorrect: The number 7 is prime, since it isn't divisible by $2, 3, 4, 5$, or 6.
  - Correct: The number 7 is prime, since it is not divisible by $2, 3, 4, 5$, or 6.

The word "precise" is not a verb in English.

- Incorrect: We now precise the connection between α and β.
- Correct: We now make the connection between α and β more precise.

Use the word "expansion", not "development" in a mathematical description.

- Incorrect: Let $[x_0, x_1, \ldots]$ be a continued fraction development of x.
- Correct: Let $[x_0, x_1, \ldots]$ be a continued fraction expansion of x.

Use "associate with", not "associate to".

- Incorrect: We now associate x to y.
- Correct: We now associate x with y.

Use "root" for equations, and "zero" for polynomials.

- Incorrect: Let α be the positive root of $x^2 - x - 1$.
- Correct: Let α be the positive zero of $x^2 - x - 1$.
- Correct: Let α be the positive root of $x^2 - x - 1 = 0$.

Use the term "pair", not "couple", to denote two objects.

- Incorrect: Let (α, β) be a couple of real numbers.
- Correct: Let (α, β) be a pair of real numbers.

Always put a comma after the Latin abbreviations "i.e." and "e.g.".

- Incorrect: Let x be an element of the set A i.e. $x \in A$.
- Correct: Let x be an element of the set A, i.e., $x \in A$.

Avoid a sentence that expresses two thoughts in a single phrase. Fix it by separating into two or more sentences, or by connecting with a semi-colon or a conjunction such as "and".

- Incorrect: Let A be a finite set, 2^A denote the family of all subsets of A.
- Correct: Let A be a finite set, and let 2^A denote the family of all subsets of A.

Avoid beginning sentences or phrases with mathematical notation or acronyms:

- Incorrect: $X(t, \phi)$ maps a random variable ϕ to a random signal.
- Correct: The function $X(t, \phi)$ maps a random variable ϕ to a random signal.

Avoid treating citation numbers as objects of prepositions. Treat them syntactically as footnotes:

- Incorrect: In [7] it is proved that $S_X(\omega)$ is the Fourier transform of $R_X(\tau)$.
- Incorrect: The article [8] proves that $S_X(\omega)$ is the Fourier transform of $R_X(\tau)$.

- Correct: Wiener proved that $S_X(\omega)$ is the Fourier transform of $R_X(\tau)$ [9].

Words like "notation" and "information" are mass nouns in English, and as such, rarely appear in the plural.

- Incorrect: One now introduces some definitions and notations.

- Correct: One now introduces some definitions and notation.

Colons should not immediately follow verbs:

- Incorrect: The resulting equation is: $x = y^2$.
- Correct: The resulting equation is $x = y^2$.

2.10.3 Common LaTeX Errors

This section lists a few of the common errors made when using the text formatter LaTeX to prepare papers.

- Blackboard Bold:
    - For blackboard bold symbols, such as, $\mathbb{Z}$, $\mathbb{Q}$, $\mathbb{R}$, $\mathbb{C}$, the authors should use \mathbb Z, for example. One may need to include the command \usepackage{amssymb}.
- Variables:
    - Variables such as x, y, n, etc., should appear in the italic font. This will occur automatically if one remembers to enclose the equations (even references to a single variable) in dollar signs or double-dollar signs, or use a latex equation environment.
        * Incorrect: Let n be the number of integers in the list.
        * Correct: Let n be the number of integers in the list.
- Floor and Ceiling:
    - Use the LaTeX commands \lfloor, \rfloor and \lceil, \rceil, not square brackets, when using these integer functions.
- Min and Max:
    - Use the LaTeX commands \min and \max when using these functions.

- GCD and LCM:
  - Use the LaTeX command \gcd for greatest common divisor (GCD). Do not write (a, b) for the GCD of a and b; write $\gcd(a, b)$ instead. For LCM, the command has to be defined so that it appears in the roman font.

- Parentheses:
  - Use parentheses to group objects, not square brackets or braces.
  - It is possible to obtain different sizes of parentheses using, for example, \bigl(and \bigr).

- Mod:
  - Differentiate between the use of "mod" as a function of two arguments, mapping $a \bmod b$ to the least non-negative residue of a modulo b, and "mod" as an equivalence relation.
  - For the first, use the LaTeX command \bmod.
  - For the second, use the LaTeX command \pmod for displayed equations.
  - For in-line equations write $x \equiv a$ (mod b), which typesets as
  $$x \equiv a \pmod b.$$

- Quotation marks:
  - Do not enclose words in ordinary quotation marks "this way"
  - This results in the following incorrect output:

    ”this way”
  - Instead, use the left-quote and right-quote symbols, or the grave and acute accents, ''this way'', which gives the correct:

    "this way"

- Proper use of \ldots and \cdots.
  - Use \ldots and \cdots properly.
  - Use \ldots, for example, if the items on either side are commas.
  - Use \cdots, for example, if the items on either side are alphabet symbols.

- For example, when referring to a product with several terms:
  - Incorrect: Consider the product $x_1 x_2 \ldots x_n$. (\ldots)

- – Correct: Consider the product $x_1 x_2 \cdots x_n$. (\cdots)
- For example, when referring to a sequence with several terms:
  - – Incorrect: Consider the sequence $x_1, x_2, \cdots, x_n$. (\cdots)
  - – Correct: Consider the sequence $x_1, x_2, \ldots, x_n$. (\ldots)
- Proper punctuation of case statements, when using braces:
  - – Punctuate case statements as follows:

$$f(x) = \begin{cases} 1, & \text{if } x \text{ is rational}; \\ 0, & \text{otherwise}. \end{cases}$$

- Definitions:
  - – Terms that are being defined can be in a special font, such as *italic* or *slant*.
  - – For example: A *rectangle* is a multi-dimensional Borel interval.
- Avoid introducing new terms and notation if there are accepted equivalents in the scientific community.
  - – For example, for the power spectral density of a stochastic signal $X(t)$ one should use the notation $S_X(\omega) = \int_{-\infty}^{\infty} R_X(\tau) e^{-j\omega\tau} d\tau$.
- Theorems:
  - – Use the \begin{theorem} ... and \end{theorem} environments for theorems, lemmas and propositions.
  - – Theorems should be numbered, because they are supposed to be referenced.
  - – It is more convenient to refer to theorems using labels, instead of using fixed references.
  - – To obtain the appropriate definitions, use the \usepackage{amsthm} command.
- Definitions, Examples, and Remarks:
  - – All definitions, examples, and remarks should be stated in the roman font, except for mathematical symbols. One can use the following code as an example:
    \theoremstyle{definition}.
    \newtheorem{defn}{Definition}.

- Proofs:
  - Use the commands \begin{proof} and \end{proof} to delimit proofs.
  - They are available in the amsthm package.
- Tables
  - Tables should be centered on the page, using the center environment.
  - Each table must have a number and a caption that explains its contents.
- Abstract:
  - Every paper must have a short abstract of 50 to 200 words, that usually comes at the beginning of the article, right after the title and author.
  - The abstract should be free of symbols whenever possible, and should not contain citations to the bibliography.
  - When referring to other work in the abstract, one can refer to author's names, but avoid mentioning years, journal names, or other information.
- Citations:
  - Use citations syntactically such as footnotes, not as objects of prepositions.
  - Avoid saying things like "In [1] we find the following result".
  - Instead, say "Lopes [2] proved the following result".
  - Use the LaTeX command \cite; do not write the actual number of the reference to the bibliography, because there is a chance it will change in the next revision.
- When listing citations, if the author has two initials, be sure to place a space between the two initials.
  - Incorrect: C.E. Shannon.
  - Correct: C. E. Shannon.
- Pay careful attention to punctuation and the use of roman, *italic*, and **boldface** fonts.
- Notice that page ranges should be separated by two hyphens in LaTeX write 123--145, not 123-145.

2.11 Bibliography Using LaTeX

- The best way to generate a bibliography at the end of the manuscript is to use the command `bibtex` for the `manuscript.tex`:

  - `latex manuscript` – first pass to create the `dvi` file.
  - `bibtex manuscript` – creates the `bbl` file.
  - `latex manuscript` – second pass to create the `dvi` file.
  - `dvips manuscript` – creates the `ps` file from the `dvi` file.
  - `ps2pdf manuscript` – creates the `pdf` file from the `ps` file.

- `@book{proakis.90,}`
  `author="John G. Proakis",`
  `title="Digital Communications",`
  `publisher="McGraw-Hill Book Company",`
  `address="New York",`
  `year="1990".`

- `@article{zadeh.65,}`
  `author="Lotfi Asker Zadeh",`
  `title="Fuzzy Sets",`
  `journal="Information and Control",`
  `volume="8",`
  `number="3",`
  `year="1965",`
  `pages="338--353",`
  `note="The very first article on fuzzy sets".`

- `@inproceedings{renyi.60.2,}`
  `author="Alfréd Rényi",`
  `title="On Measures of Entropy and Information",`
  `booktitle="Proceedings of the 4th Berkeley Symposium on Mathematics,`
  `Statistics and Probability",`
  `publisher="University of California Press",`
  `address="Berkeley, USA",`
  `volume="1",`
  `pages="547--561",`
  `year="1961".`

2.12 Chapter Summary

This chapter covers the mathematical style in English, ranging from reading to writing, while also dealing with the combination of numbers, words, formulas and theorems. The mathematical style has the objective of enabling the readers to understand mathematical concepts, use the right notation, and write properly.

The reader can expect to learn how to handle the most important parts of the text, how to manage the comments and tips, how to prepare books for publication and a helpful guide for Linux and Latex usage.

Mathematics is an important and unique characteristic of human thought. The purpose of this chapter is to help with writing scientific texts in English.

2.13 Reference Material

- *Writing Mathematics Well* by Leonard Gillman. Mathematical Association of America.
    - Gillman's book refers to the three previous classics in the field:
        * An article by Harley Flanders, *Amer. Math. Monthly*, 1971, pp. 1–10.
        * Another by R. P. Boas in the same journal, 1981, pp. 727–731.
        * There's also a nice booklet called *How to Write Mathematics*, published by the American Mathematical Society in 1973, especially the essay by Paul R. Halmos on pp. 19–48.
- *Mathematical Writing*, by Donald E. Knuth, Tracy Larrabee, Paul M. Roberts. Mathematical Association of America, 1989.
    - This book is based on a course given by Donald Knuth at Stanford University. It discusses technical writing and presentation of mathematics and computer science, including preparation of theses, papers, and books.
- MIT Guide to Science and Engineering Communication, J. G. Paradis and M. L. Zimmerman. MIT, Cambridge, 2002.
    - Discusses all types of technical communication and includes a list of 27 guidelines for style and usage.
- *English for Electrical Engineers*, by James MacAllister and Giorgio Madama. Longman, 1976.

- The book presents and analyzes interesting texts, in increasing order of difficulty, along with language tips and grammar.

- *A History of Mathematics*, by Carl B. Boyer and Uta C. Merzbach, Wiley, 2011.

  - The text takes the style of a history book, and often focuses on the people and on the time period, commenting on political and cultural aspects. It is enjoyable to read.

- *Encyclopaedia of Mathematics*, published by Kluwer Academic Publishers, 2002. Also in the Internet: www.encyclopediaofmath.org.

The TEX and LATEX Reference Books

- The TEXbook (Computers & Typesetting), by Donald E. Knuth. Addison-Wesley Professional. 1984, New York, United States.

  - This is the reference book on TEX.

- LATEX A Document Preparation System (2nd Edition), by Leslie Lamport Addison-Wesley Series on Tools and Techniques for Computer T, 1994, New York, United States.

  - This is the reference book on LATEX.

3

Technical Style

"I very rarely think in words at all. A thought comes, and I may try to express in words afterwards."
Albert Einstein

3.1 Introduction

Technical writing should be straightforward, easy to read and understand. It must also be efficient and clear, non-personal, and should achieve a balance between being thorough enough to explain what is presented, without being prolix.

To pursue those objectives, the author should always be specific, especially when using comparatives and superlatives, it is important to put statements in a positive form, use numerical values, avoid ambiguity, and only compare values when they have the same units.

3.2 Definition

The technical style has the objective of enabling readers to use a technology or understand a process or concept. Most technical documents use an objective style, because the subject matter is the center of the attention.

In addition, the technical style is direct and utilitarian, emphasizing exactness and clarity, rather than refinement and allusiveness, but it does not disregards elegance. The elegance of a mathematical proof relies on scientific precision, neatness, and simplicity.

3.3 Overview of the Chapter

This chapter focus on technical and scientific writing, including the process of preparation of a document, the organization of an article, and how to coordinate figures and tables.

The reader will learn the intricacies of writing a manuscript, that mostly deals with revision and style, word choice and usage, grammar and punctuation. Some basic problems in technical writing are discussed, along with useful writing tips.

3.4 Technical Writing

The author generally wants to communicate scientific or technical information, in a precise manner, to a specialized audience, therefore the main objectives in technical writing are:

- make complex technical information understandable.
- make it easy for the reader to read and extract information.
- achieve clarity, conciseness, and coherence.

3.5 Precise Writing

Precise technical and scientific writing is an important skill that can be learned and mastered, but it takes time and hard work, and a good deal of reading.

The process requires the use of references, which generally include:

- Grammars,
- Dictionaries,
- Thesaurus,
- Handbooks on writing.

3.6 English as a Second Language

English as a language has great reach and influence. It is taught all over the world, and substituted French, the *lingua franca* of the XIX Century, and Latin, which preceded it, and ruled for centuries, as the fundamental means of communicating scientific work.

The main problems associated with choosing English as an international standard are:

- Those who learn English as a second language face special challenges.
- Each language has its own rules and characteristics, although they usually share general rules of grammar.

- There is a natural tendency to carry the problems with a native language over into English.

Some of the common usage problems are:

- Transitive verbs: This technique allows to ...
- Nonexistent words: modelizations.
- Missing articles: a, an, the.
- Misused pronouns: It means that, instead of That means that.

It is possible to learn how to cope with those problems using some classical references:

- Handbook of Technical Writing.
- Mayfield Handbook Online Writing Lab.
- An Outline of Scientific Writing.

3.7 Reader's Approach to Read an Article

It is a fact that most readers do not read the whole article. At first, the reader browses the article, or report, to obtain basic information. The typical order in which they will read the article:

- Title and author list.
- Abstract and keywords.
- Figures and tables.
- Browse text and section headings.
- Conclusion.
- Equations.
- Portions of main text in more detail.

Therefore, it is recommended that the writer makes sure that the main elements of the list are well-prepared.

3.8 Title of the Document

The title is the most visible part of article, and it must convey information on the article contents. The main goals for the title are:

- it must be informative about what is in the paper.
- no longer than ten words.
- it is distinctive.

And it is important to realize that the title is:

- not a replacement for the abstract,
- not a sentence, usually.

3.9 Title of the Document

For the title:

- do not start with an article.
- avoid all but the best-known acronyms.

3.10 Abstract and Keywords

3.10.1 Abstract

Researchers usually focus on the methods, results and discussion when creating a document, little thought is given on the title and the abstract, while even less thought is spent on the keywords, which are many times typed only at the moment of submission, ironically, those three elements can be key to the success of the article.

The abstract must be:

- concise, clear, and an informative summary of work in paper.
- a single paragraph not very long (less than 250 words).
- without lengthy background.
- readable and impressive.

Keywords or citation indices should be selected very carefully, because the researchers will first search databases for keywords.

3.11 Figures

Figures and their captions help describe the contents of the technical or scientific document, therefore, they should describe results independently from text, the captions must be clear and use an adequate type size.

The writer must anticipate how graphs and images will appear in a published article, by:

- adjusting their dimensions for one or two columns,
- and making sure that:
    - the thickness of lines and axes are correct,

- the symbols and fonts have the adequate dimensions,
- everything fits the size and proportion of the final graph.

Regarding the use of color and other details, it is important to note that:

- one should use color to distinguish lines only if published paper will be in color.
- one should use solid, dashed, and dotted lines, and various data symbols.
- caption should describe the figure and provide link to text.

3.12 The Writing Process

The writing process involves planning, organization, preparing the first draft, and the revision, which is a recursive process.

Planning is:

- identifying an objective.
- finding the right audience,
- devising the scope of the paper.

Organization is:

- the logical development of the text.
- is preparation of an outline.

First draft is:

- writing a rough draft of the paper.
- revising the document to refine it.

Revision:

- The goal of revision is completeness, accuracy, and coherence.
- The document must be edited for style, word choice, and grammar.

3.13 Planning

It is finding the best approach, and starting the writing process with a plan.

The purpose of the article:

- is to solve a problem.
- is to convey new information.
- is to express a point of view.
- is to persuade reader for something.

Identify the audience:

- To whom are you telling the story?
- Why would someone want to read your article?
- What is the expected level of expertise?

Determine the scope of the presentation:

- Depends on the purpose and the audience.

3.14 Outline

Before beginning to write, create an outline.

- It is used as the skeleton for the manuscript.
    - To provide organization and structure.
    - To establish overall logic of presentation.
- Try to include every topic you want to mention.

The following techniques can help start:

- Define the essence of the message in a few main ideas.
- Write down key points first, then take care of secondary ideas.
- Give an informal talk to friends or colleagues, to obtain positive criticism.
- Maintain momentum – do not stop prematurely, and do not go on forever.

3.15 Writing the First Draft

Base the first draft of the outline. Turn off all the distractions and focus only on writing.

- Outline provides organization:
    - Topics and subtopics of outline become sections and subsections.
    - Paragraphs derive from subtopics and sub-subtopics.
- Skip Abstract, Introduction, and Conclusions.
    - They are easier to write after the document is ready.

After writing the first draft, let it breath, get up and go do something else, and then return to the draft, depending on the deadline and type of text that is being written, do something else for an hour (if the deadline is tight) or, if it is a report or a book, set the draft aside for a day or two, but keep thinking about it.

Do not be afraid to rewrite the first draft, as many times as possible, this way the learning process is enhanced, as well as the quality of the document. Exercise your vocabulary, by keeping a dictionary nearby. Do not repeat words in the same sentence, or in the current paragraph.

> A journalist once interviewed Ernest Hemingway, and began asking: How much rewriting do you do?
> Hemingway responded: It depends. I rewrote the ending of Farewell to Arms, the last page of it, 39 times before I was satisfied. The journalist insisted: Was there some technical problem there? What was it that had stumped you?
> Hemingway then replied: Getting the words right.
> – Ernest Hemingway, The Paris Review Interview, 1956.

Avoid procrastinating, it is the worst thing you can do, since then, you will see writing as a chore, and not a nice hobby or promising career.

Always keep several backups of the writing, in different types of media. This way, if the computer or the files present issues, the work is not lost.

3.16 Techniques for Writing the First Draft

Useful techniques for beginning to write:

- Write first draft very quickly (and roughly).
    - Do not check spelling and style at this phase.
    - Start with sections that are easy to write.
    - Write in stream-of-consciousness mode.

Writing conditions:

- Set aside blocks of time to write, perhaps an hour or two.
- Establish goal for writing in each session.
- Make sure your environment is conducive to writing.

3.17 Revision of the Document

It is important to consider that the first draft is not ready to show anyone until after the first revision. Most authors say that the secret of good writing is to review the document several times, leaving an interval between the revisions.

How to Divide the Revision Process

In general, the final preparation of a document could be separated into the following parts: large-scale revision, small-scale revision, editing and proofreading.

- Large-scale revision – This type of revision is all about the big picture, scanning the entire document for parts that could use an improvement. It might be necessary to present evidence, define terms, or add a completely new step to the reasoning. It could even come down to a complete restructure or rewrite of the paper if a new intriguing idea is discovered, or a different structure seems to be more effective than the one currently being used.
- Small-scale revision – It is a tool for when it becomes clear that a specific part of the paper is not working. It could be the introduction, or a part of the argument may seem weak. Once the problem is isolated, focus on the revision of that specific section of the paper. When the process is done, make sure the new revisions are a good fit in the context of the paper in its entirety.
- Editing – Frequently students confuse editing with revision. Those are two distinct processes. Editing is finding the small issues within the text, problems that can be fixed by simply deleting, or replacing, a word or sentence, switching the position of a paragraph etc. When editing, consider the reader, and use some empathy. Better yet, consider yourself a reader. Ask yourself if the reader will find the paper clear, concise, readable and interesting. And question how can the paper be rewritten to enhance clarity, conciseness, and the pleasure to read.
- Proofreading – When reading the final version of the paper, report or book, try to find any mistakes. The most common errors are punctuation mistakes, incorrect spelling, subject-verb agreement, common mistakes, such as, confusing its/it's, their/there etc. Read slowly, focusing on every word and every phrase of the paper. Remember to use the spell-checking feature of your word processor, or the command `ispell`, if you are a Unix or Linux user. But, the best way to improve your writing skills is to browse a dictionary or a grammar.

Revision is a Critical Step in Writing a Manuscript

- Good writing depends strongly on revision.
- It usually takes several passes through the manuscript.

Review Content and Organization

- Does it say what is meant?
  - does it include all the data, graphs, etc.?
- Is it easy to read and follow logic of presentation?
- Is it accurate, complete, and truthful?

Check for Style and Proper English

- Clarity, conciseness, and coherence.
- Sentence construction.
- Word choice and usage.
- Grammar, punctuation, and spelling.

3.18 Revision Strategies

Some useful strategies for revising a manuscript are listed in the following. But first, a useful advice: Create a new version, but keep old versions until the work is finished.

- Revise the manuscript on a computer monitor, for the initial versions.
- Print and read the document, for the final version.

The general approach to revise the manuscript includes:

- seeing the complete manuscript.
- rearranging sections and paragraphs, as a way to improve development.
- identifying what is missing and adding new text to complete the ideas.
- reading the manuscript several times, each time looking for a particular type of problem.
- making cursory notations in text or margin, during the reading, and correct later.
  - use standard proofreading marks, especially if someone else will read it.
  - use $1\frac{1}{2}$ to 2 times line spacing to allow insertion of notes and new text.

3.19 Good Technical Writing Style

3.19.1 Technical Writing

- Style is how you say things in your writing.
- Goal of technical writing is clarity, conciseness, and coherence.

- Use straightforward and simple sentence construction.
- Choose words carefully.
  - – aim at conciseness and clarity and avoid wordiness.
  - – avoid colloquialism, slang, and shoptalk.

3.20 Good Technical Writing Style

For a good technical writing style it is important to use correct word usage, and to study grammar, punctuation, and spelling.

Reading good authors is the main advice to those who wish to write correctly. Anyway, the most common problems encountered in the technical, and non-technical, texts are described in the following.

3.20.1 Produce a Text That Is Readable and Easy to Follow

- Maintain overall organization.
- Use transition elements throughout.

3.21 Transition Elements

Transition elements are crucial for keeping the reader on track. Their main purpose is to link together different parts of the article. Effective transitions are needed at all levels of the article structure.

- Article
  - – introduction connects with previous work and lays out organization.
  - – conclusion summarizes what has been presented.
- Section
  - – begin each section with short introduction to establish its relationship to previous section and the overall context.
- Paragraph
  - – use topic sentence and logical development within each paragraph.
  - – establish links between paragraphs.
- Sentence
  - – use simple and short sentences to express the main ideas.

– if necessary, use compound sentences with transition or subordinating conjunctions.

3.22 Paragraphs

The paragraph is the unit of composition. The paragraphs should not be very long or very short. An extensive paragraph is difficult to read, and understand. A reduced paragraph probably does not convey the information it should.

3.22.1 Organizing Principles

- Unity
    - Focus on a central topic sentence.
        * placed first, second, or last in paragraph.
- Development
    - Advance the topic with logical arguments.
- Coherence
    - Sentences should hang together.
    - Transition elements link sentences.
        * – connecting phrases (On the other hand, ...; Therefore, ...).
        * – repetition of keywords.

3.23 Sentences

The sentence structure should generally be simple. If a sentence can be shortened, without losing meaning, it must be reduced. Keep the sentence length moderate.

- To promote clarity and readability use subject-verb-object construction.
- Avoid complicated structure to explain complex ideas.

Prefer, when writing a technical or scientific text:

- Strong verbs.
- Active voice.
- Impersonal.
- First person, when appropriate.

Equations are part of a sentence; therefore, punctuate accordingly.

- Equation numbers are usually presented in parentheses, e.g., Equation (9).

3.24 Compound Sentences

Deviation from simple sentence construction can be helpful in technical writing, to convey the appropriate meaning to a subject.

3.24.1 Compound Sentences Can Be Used to

- Make transitions.
  - The amplitude modulator was saturated; therefore, it was operating in the non-linear region.
  - The image quality is adequate, but it is also necessary to verify the texture.
- Indicate subordination of ideas.
  - Because the input signal was strong, there was no need for an encoder.
  - The oscilloscope, which was acquired by the electrical engineering department, allowed to visualize the problem.

3.25 Verbs – Tense

3.25.1 Use Present Tense as a General Rule

- Although it may seem unnatural to write about the past in the present tense, it is usual.
- The document is always in the present of the reader.
  - There is no section in the past or future.
  - There is no section above or below.

3.25.2 Other Tenses May Be Used

- Past may be used in Introduction.
  - to refer to previous work.
- Past may be used in Discussion.

– in describing materials used.
– to refer to set up of experiment.

- Future may be used in Conclusion.

– to refer to future work.

- Do not to switch tenses.

3.26 First Person

Write in first person, but only when appropriate.

- Person indicates the writer's relation to the material presented.

– writing in first person shows direct involvement.
– writing in second or third person indicates impersonal relation.

- Use first person, singular if one author and plural for two or more.

– plural first person may be used for a single author to include reader.
– I conclude that the result is correct. We can conclude that there is a problem with the model.

- Use first person when writing about.

– your choices, opinions, expectations.
– your measurements, calculations, conclusions.

Writing in first person tends to promote the active voice, but it is not generally used in technical documents. Although, it is tolerated in presentations.

- Passive: The results are calculated using the Monte Carlo method.
- Active: We calculate our results using the Monte Carlo method.

3.27 Word Choice

It is important to chose the words carefully to convey the precise meaning. Good writers are also good readers, and they usually pick the classics, to maintain a certain level of quality in writing.

- Pick powerful words with definite meanings and, once more, avoid:

– the use of words that may cause ambiguity,
– ornate or erudite words, if not in a specific context,
– wordiness and redundancy,

- informal English usage,
- idioms, unnecessary jargon, slang.

3.28 Measurement Units

Measurement units are usually abbreviated, as in the examples that follow:

- mm – millimeter; length.
- km – kilometer; length.
- s – second; time.
- Hz – hertz $= s^{-1}$; frequency.
- MHz – megahertz; frequency.
- pt. – point; length in type setting units.

Generally, it is necessary to include a small space between the number and the unit. Do not italicize a unit, it is always written in roman.

- Incorrect: 2.47 *mm*; Incorrect: 2.47mm; Correct: 2.47 mm.
- Incorrect: 6pts.; Incorrect: 6 *pt.*; Correct: 6 pt.
- But: 54°C.

3.29 Common Problems in Writing

Some of the common problems in technical writing are related to the use of passive voice, nominalization, wordiness, missing articles and commas, compound modifiers, treating countable nouns as uncountable, the use of transitive verbs without a direct object, inappropriate use of words, jargon, acronyms, and punctuation.

Passive Voice

- active voice improves clarity.
  - Passive: It was proved by Kolmogorov, in 1931, that ...
  - Active: In 1931, Kolmogorov proved that ...

Nominalization (weak verb + noun)

- Instead, use a strong verb.
  - Incorrect: We perform a calculation using Equation (8) to obtain the results shown in Figure 2.

– Correct: We calculate the results shown in Figure 2 using Equation (8).

Wordiness

- Eliminate unnecessary words to achieve conciseness.
- Watch out for wordy clichés, e.g., *for the reason that* change to *because*.

Comma Missing after Introductory Phrase or Clause

- Incorrect: To test our hypothesis we calculate the probability.
- Correct: To test our hypothesis, we calculate the probability.

Compound Modifiers (adjectives) without Hyphens

- Use hyphens to connect modifying words that go together.
  - Incorrect: Therefore, the high frequency components are attenuated.
  - Correct: Therefore, the high-frequency components are attenuated.

Missing or Inappropriate Articles (a, an, the)

- Incorrect: Algorithm permits the computation of definite integral. (articles missing)
- Correct: The algorithm permits the computation of a definite integral.

Treating Countable Nouns as Uncountable

- Incorrect: less problems ...
- Correct: fewer problems ...
- Incorrect: so much artifacts ...
- Correct: so many artifacts ...

Transitive Verbs without a Direct Object

- Incorrect: The algorithm allows to calculate the integral. (object missing)
- Correct: The algorithm allows one to calculate the integral.

Inappropriate Use of Words

- In order to – should not generally be used, except to avoid ambiguity.
  - Incorrect: In order to modulate the signal.
  - Correct: To modulate the signal.
- Which, that, who.
  - use that before a restrictive phrase (without comma).
    * The approach that provides the higher gain.
  - use which to begin a nonrestrictive phrase, with comma before and after.
    * The approach, which was adopted from Papoulis, proved to work well.
  - use who when referring to a person or people.
    * People who follow Papoulis' approach.
- Due to – do not use in place of because of ...
  - Incorrect: The computer failed due to ...
  - Correct: The computer failed because of ...
- Data is a plural countable noun, especially in technical writing; also spectra, criteria, phenomena, momenta, radii, ...
  - Incorrect: The memory data is lost ...
  - Correct: The memory data are ...
- This, at beginning of sentence with no following noun, often indicates ambiguous reference.
  - Incorrect: This means that ...
  - Correct: This result means that ...

Inappropriate Use of Jargon

Jargon and slang are two special types of language varieties, but there is a difference between them. Jargon is a terminology, from a specialized vocabulary, that is used in relation to a profession, to a certain activity, to occupational or recreational groups, or to an event.

Slang is the informal use of words and expressions that are not considered standard in the English language. The use of slang indicates a desire for social distance to language speakers that are outside the group and their mainstream values.

An appropriate use of jargon depends on the expertise of the intended audience. In general, it is better to avoid jargon, and slang is not permitted in a technical document.

- Incorrect: The microphone is producing a loud noise.
- Correct: The microphone is producing a loud noise.

Too Many Acronyms

The acronyms should be defined at first use, with no exceptions.

- Incorrect: The CDF approaches one, when the random variable goes to infinite.
- Correct: The cumulative distribution function (CDF) approaches one, when the random variable goes to infinite.

Inappropriate Use of Punctuation

- Correct punctuation enhances readability.
  - (,) comma – pauses the flow of a sentence to prevent ambiguity (e.g. series, introductory phrase, nonessential phrase).
  - (:) colon – initiates series of terms, equations.
  - (;) semicolon – initiates independent clause.
  - (–) dash – sets off phrases with emphasis.
  - ( ) parenthesis – encloses nonessential words and phrases.

3.30 How to Prepare a Summary

Organization of Material Is Important
Before starting the writing, it is important to obtain all the necessary material, including related articles, books, reports, manuals, either in print or as computer files.

Good Technical Writing Style Is Learned by
- Reading well-written journal articles and books by good authors.
- Paying attention to the details.
- Using writing guides and dictionaries, especially when in doubt.
- Having your writing critically edited by a technical editor or colleagues.

Find the Best Approach to Writing

It is necessary to find the approach to writing that works best for a particular author. Some authors write from scratch, others do research first. Some authors write in a pipeline mode, one chapter at a time, others write the chapters in parallel.

Revise the Writing

Writing well involves revisions, and the number of revisions and time needed greatly depend on the type of document and the writer's skill. Usually, spending too much time on the review can lead to boredom, and spending a short time can result in errors getting through to the final draft.

Rushing to get a product in the market or a paper published, can lead to a poorly written report, an incomplete service manual, an article full of mistakes, prone to be rejected.

Reviewing the draft, and focusing on one error at a time, might be an effective way to catch mistakes, but certainly not the most efficient. But the author must devote enough time to review it, and then decide that it is good enough, when feeling good about the writing.

It is important to get a colleague, or hire a proofreader, to review the work to identify errors. An alternative is to perform a thorough review and then set it aside for a day or two. Then, return to it, and the mistakes will certainly appear. Repeat the process until the work is done.

3.31 Chapter Summary

Technical, as well as, scientific writing style is presented, including the process of preparation of a document, including the planning, the outline, the preparation of the first draft, organization principles and measurement units.

The chapter also deals with precise writing, with revision and style, word choice and usage, grammar and punctuation. Some usual problems in technical writing are discussed.

3.32 Reference Material

Recommended Guides

- Mayfield Handbook of Technical and Scientific Writing, L. C. Perelman, E. Barrett; and J. Paradis; `mit.imoat.net/handbook/home.htm`.

- recommended; complete guide to technical writing from MIT; concise explanation of most aspects of technical writing; ESL pointers.
- Online Writing Lab (OWL); `owl.english.purdue.edu/handouts`
  - guide to effective writing at college level; grammar and punctuation with exercises; English as a Second Language (ESL).
- Supporting material for book Handbook of Technical Writing, Alred et al.; `bcs.bedfordstmartins.com/alredtech`.
- Grammar, Punctuation, and Capitalization: A Handbook for Technical Writers and Editors; `stipo.larc.nasa.gov/sp7084`.
  - NASA Report; hypertext or PDF; rules for technical writing.
- AIP Style Manual; `public.lanl.gov/kmh/AIP_Style_4thed.html`.
  - American Institute of Physics gives stylistic guidance, especially relevant to physics articles.

Online Writing References

- Merriam-Webster Dictionary and Thesaurus, can be found at `www.m-w.com`.
  - Usable, gives etymology and pronunciation of words.
- Wordsmyth Dictionary and Thesaurus, can be found at `www.wordsmyth.net`.
  - Very usable, although definitions are brief, it identifies parts of speech.

Bartleby Classic Online Books `www.bartleby.com`

- A good collection of writers' aids, that includes the American Heritage Dictionary, Roget's Thesaurus, quotations, and more:
  - American Heritage Dictionary of the English Language `www.bartleby.com/61`.
  - The Elements of Style `www.bartleby.com/141`, classic handbook, written by William Strunk in 1918
  - The King's English `www.bartleby.com/116`, by H. W. Fowler (1908), another classic.

Reference Books

- Handbook of Technical Writing, G. J. Alred, C. T. Brusaw, and W. E. Oliu (St. Martin's, New York, 2003).
    - Highly recommended; complete handbook on technical writing; entries arranged in alphabetical order; excellent index; English as a second language guidance; includes succinct guide to the writing process.
    - Supporting material at `bcs.bedfordstmartins.com/alredtech`.
- An Outline of Scientific Writing: For Researchers With English as a Foreign Language, J. T. Yang (World Scientific, 1995).
    - May be especially useful to English as a second language writers.
- MIT Guide to Science and Engineering Communication, J. G. Paradis and M. L. Zimmerman (MIT, Cambridge, 2002).
    - Discusses all types of technical communication and includes a list of 27 guidelines for style and usage.

4

How to Read a Document

"Information is not knowledge."
Albert Einstein

4.1 Objectives of Reading

The objective of the reader, which could be entertainment, information or understanding, usually determines the way to approach what to read. The approach is different, if the text is a document, a report, an article, or a book.

The reading effectiveness is then determined by the amount of effort and skill put into the process, which also depends on the subject to be read. For most people, Mathematics, Physics and Engineering texts demand additional effort to read.

A trained mind can obtain more information from the text. The distinction between instruction and discovery, or aided and unaided discovery, is equivalent to learn with and without a teacher.

According to Mortimer Adler and Charles van Doren, there are four levels of reading, in increasing order of difficulty, the higher levels include the lower ones, cumulatively.

4.1.1 Levels of Reading

The first level is Elementary Reading, but it could be also called rudimentary, basic or initial reading. Any of those terms suggest that one masters that level when one passes from illiteracy to, at least, beginning literacy.

The level implies that the reader masters the rudiments of the art of reading, receives basic training in reading and acquires initial reading skills. At that level of reading, the main question is "What does the sentence say?".

The first effort is to identify the actual words, and after recognizing them individually, begin to understand them and perceive what they mean.

The second level of reading is Inspectional Reading, also called systematically skimming or pre-reading, which is characterized by its special emphasis on time.

When reading at that level, the reader is allowed to set a time to complete an assigned amount of reading. Therefore, the goal is to obtain the most out of a book within a given time, usually too short to get out of the book everything that can be obtained.

When reading at this level, the aim is to examine the surface of the book and to learn everything that the surface itself can teach the reader.

The questions asked at this level are "What is the book about?", "What is the structure of the book?" or "What are its parts?".

The reader can obtain information from the title, contents, preface, index, appendices, and editor's blurb on the book jacket, if available.

The third level of reading is Analytical Reading, which is a more systematic activity, also called thorough, complete or good reading, or the best reading that can be done given unlimited time.

The analytical reader is active and must ask many, and organized, questions of what to read.

Analytical reading is for the sake of understanding, and hardly ever necessary if the aim is simply entertainment or information. It is related to the process of bringing the mind, with the aid of the book, from a condition of understanding less, to the state of understanding more. It is necessary, for instance, to read a profound book or a thesis.

- The first stage of analytical reading – Rules to find what a book is about.
  This stage of this level of reading is concerned with understanding the structure and purpose of the book. The reader establishes the basic topic and type of the book being read, to anticipate the contents and comprehend the book from the beginning, and distinguish between practical and theoretical books, as well to determine the field of study that the book addresses, and note any divisions in the book, not restricted to the divisions defined by table of contents. Finally, the reader must find out what problems the author is trying to solve.
- The second stage of analytical reading – Rules to interpret a book's content.
  This stage of analytical reading involves constructing the author's arguments. This requires the reader to note and understand any special phrases and terms that the author uses. Then, the reader should find and

work to understand each proposition that the author advances, as well as the author's support for those propositions.

- The third stage of analytical reading – Rules to criticize a book as a communication of knowledge.

In this stage, the reader must criticize the book. After the author's propositions and arguments are understood, the reader is now able to judge the book's merit and accuracy. Of course, the reader should not begin criticism until completing an outline and interpreting the book. Good reasons must be presented for any critical judgment made. Some criteria for points of criticism include to show wherein the author is uninformed, misinformed, illogical or incomplete.

4.1.2 The Highest Levels of Reading

The fourth and highest level is Syntopical Reading, also called comparative reading, which puts heavy demands on the reader even if the materials are relatively easy. The reader must be ready to compare different texts, and extract the most of the reading.

The expression was coined by Mortimer Adler in "How to Read a Book", and was derived from the term Syntopicon®, a two-volume topical guide published as part of the Encyclopedia Britannica's collection Great Books of the Western World series. It refers to a type of analysis in which different works are compared and contrasted.

When performing a comparative reading, it is important to read many books and place them in relation to one another, and to a subject about which they all revolve. With the help of the books read, the reader is able to construct an analysis of the subject that may not be in any of the books, which is akin to the type of creative research required to write a manuscript or thesis. It is considered the most rewarding of all reading activities.

According to Mortimer Adler and Charles van Doren, there are two main stages of comparative reading: one is the preparation, and the other is the comparison reading proper.

- Survey the field as a preparation for comparative reading.

  1. Create a tentative bibliography of the subject, using library catalogues, advisers, and bibliographies in books.
  2. Inspect all of the books on the preliminary bibliography to define which are related to the subject, and also to acquire a clearer idea of the theme.

- Comparative reading of the collected bibliography in the first stage.
    1. Inspect the books to find relevant passages.
    2. Bring the authors to terms by constructing a neutral terminology of the subject.
    3. Establish a set of neutral propositions for all of the authors by framing a set of questions to which most of the authors can be interpreted as giving answers, whether they actually treat the questions explicitly or not.
    Finding the answers in the author's text to one's own question.
    4. Define the issues, both major or minor, by ranging the opposing answers of authors to the various questions on each side of an issue.
    5. Analyze the discussion by ordering the questions and issues in a way to throw maximum light on the subject. More general issues should precede less general ones, and relations among issues should be clearly indicated.

4.2 How to Read Faster

Most persons read, on average, 200 words per minute. People who learned to read fast can read five times more than that, approximately a thousand words. In the following there are some tips on how this can be done. It is not necessary to spend time and money on training sessions or computer programs.

First, do not try to read everything quickly! While reading a good book or needing to understand an entire document (a technical report, for instance), take some time to read them correctly. Reading fast is not only about speed, it is about making the reading more efficient. The intention is to help the reader identify the parts of a text which need more attention, and oversee the rest.

4.2.1 Pre-Reading Is the First Step

Before starting on a text, a reader must decide how fast to read it. It is a good practice to skim over the whole text, checking the titles, the mathematical notation, the equations and formulas, the items and the boldfaced words.

This will allow the identification of the main material. After that, go back and re-read the truly important parts at the right pace, paying attention to the important passages and deductions.

4.2.2 Read the Text in Blocks

Most students were taught to read one word at a time, line by line. However, it is possible to absorb blocks of text while reading, instead of focusing on one word at a time. If the reader can do that, the comprehension of the material can be enhanced.

For this, it is necessary to keep the text a bit further from the eyes, then one would normally do while reading, and try to relax the face and the eyes. The idea is to train the eyes to observe more words at once: lines or entire sentences instead of individual words.

It is a good practice to focus the middle of each line, instead of starting from the start of the line, and allow to comprehend the meaning of the text on the line and advance to the next one. When that can be done, try to focus the middle of a paragraph, and then an even larger block of text, even an entire page.

The reader will be surprised with the volume of information that can be absorbed. Once again, the objective is not to understand the text profoundly, but to scan it quickly and identify what it is really needed to focus on.

4.2.3 Subvocalization Is a Problem

This is the main cause of slower reading. While reading a person pronounces each word, not aloud, but inside the head. Most people do not even notice they do this. Read this sentence, and pay attention if the words are said under the breath, inside the head or even moving the lips, during the reading. If the reader can notice it, it is possible to consciously turn off the voice.

It is a fact is that people talk at an average speed of 120 words per minute, therefore, vocalization, the act of saying the words while reading, indicates that the reader will only be able to read at talking speed.

A technique to turn off the voice in the head, is to try humming while reading or counting to four repeatedly. In other words, the secret is to keep the mouth, tongue and breathing busy with simple repetitive sounds while the eyes advance effortlessly through the page. This technique really works, and it is possible to test it reading this paragraph again while counting in the head.

4.2.4 Read Only Once

Re-reading is a habit, and can be eliminated. Unless it is really important to recall the words, or when reading a complex document, a mathematical or engineering thesis, for instance, the reader will probably understand the text on the first read.

Therefore, it is not advisable to re-read the material. A pen or a finger can be used to hover over words already read, and the eye will follow the pointer. Or better yet, cover the lines during the reading with a bookmark or piece of paper, so that the eyes can not go back. The speed can be increased when the reader begins to feel comfortable.

Some people think that, to be a great entrepreneur, they need to accomplish several tasks at once. But that can greatly reduce the focus. Observe someone trying to read in a noisy place, and it is almost certain that such person is mouthing the words in the head, and also re-reading to focus.

Therefore, to obtain the maximum out of a document, report, book or article, it is necessary to turn off the cell phone, player, radio or television set, and go to a room where no one is talking, and no distractions are found.

4.3 Chapter Summary

This chapter has the objective of explaining the steps to obtain higher focusing capability, faster and better reading, and time management while also retaining the essential information in the most efficient manner.

The first level is Elementary Reading, also called rudimentary, basic or initial reading. The level implies that the reader masters the rudiments of the art of reading, receives basic training in reading and acquires initial reading skills.

The second level of reading is Inspectional Reading, also called systematically skimming or pre-reading, which is characterized by its special emphasis on time. When reading at that level, the reader is allowed to set a time to complete an assigned amount of reading.

The third level of reading is Analytical Reading, which is a more systematic activity, also called thorough, complete or good reading, or the best reading that can be done given unlimited time.

The fourth and highest level is Comparative Reading, which puts heavy demands on the reader even if the materials are relatively easy. The reader must be ready to compare different texts, and extract the most of the reading.

The chapter also gives a few tips on how to read fast, which is not only about speed, but also about making the reading more efficient. While reading a good book or needing to understand an entire document, take some time to read them correctly.

- Pre-reading is the first step – Before starting on a text, a reader must decide how fast to read it.

- Read the text in blocks – It is possible to absorb blocks of text while reading, instead of focusing on one word at a time.
- Subvocalization is a problem – This is the main cause of slower reading.
- Read only once – Re-reading is a habit, and can be eliminated.

The intention of the final part of the chapter is to help the reader identify the parts of a text which need more attention and survey the rest.

4.4 Reference Material

- How to Read a Book, Mortimer Adler and Charles van Doren. MJF Books, New York, USA, 1972. ISBN 1-56731-010-9.
    - A classic book describing the four stages of reading, includes a list of relevant books.
- How to Read Better & Faster, Paperback, by Norman Lewis. Goyal Publishers & Distributors Pvt. Ltd, 2006.
    - An interesting book that teaches reading techniques, and describes the main steps to read faster.
- Handbook of Technical Writing, Gerald J. Alred, Charles T. Brusaw, Walter. E. Oliu. St. Martin's Press, New York, 2006.
    - Highly recommended; complete handbook on technical writing; entries arranged in alphabetical order; excellent index; English as a second language guidance; includes succinct guide to the writing process. Supporting material at `bcs.bedfordstmartins.com/-alredtech`
- An Outline of Scientific Writing: For Researchers With English as a Foreign Language, J. T. Yang. World Scientific, 1995.
    - May be especially useful to English as a second language writers.
- MIT Guide to Science and Engineering Communication, J. G. Paradis and M. L. Zimmerman. MIT, Cambridge, 2002.
    - Discusses all types of technical communication and includes a list of 27 guidelines for style and usage.
- *English for Electrical Engineers*, by James MacAllister and Giorgio Madama. Longman, 1976.
    - The book presents and analyzes interesting texts, in increasing order of difficulty, along with language tips and grammar.

5

Stylistic Problems in English

"Everything should be made as simple as possible, but not simpler."
Albert Einstein

5.1 Definition

In Plutarch's (ca. 46–120 AD) "Parallel Lives", when Pompey was about to set sail, there was a violent storm at sea, and the ship-captains hesitated to leave. But he led the way on board and ordered them to weigh anchor, crying out loud: "To sail is necessary, to live is not."

Writing a technical text is somehow different from talking about the subject. When explaining a certain subject, the instructor usually has at her disposal a blackboard, a flip-chart, or a projector, to help her detail and illustrate the important points.

But, when a professional is reading a technical manuscript, the only source of information is the document itself. Therefore, it has to be well and clearly written, accurate and devoid of mistakes. Thinking about Plutarch's work, one can summarize the previous two paragraphs in one sentence: "To write is precise, to speak is not."

Because style is a convention with respect to spelling, punctuation, capitalization, and typographic arrangement and display followed in writing or printing, it is important to avoid come common problems that appear during the writing of a technical or literary text. Therefore, the writer has to follow the rules to be properly understood.

5.2 Problems with Verbs

5.2.1 Verbs – Tense

Use consistent tense (present, usually, unless reporting results achieved in earlier papers)

None can take either singular or plural verbs, depending on the intended meaning or taste.

- Both "none of these mistakes are common" and "none of these mistakes is common" are correct.

5.2.2 Passive Tense

Avoid use of passive tense.

- Example: "In each message, a subject is included." reads better and shorter as "Each message includes a subject".

5.2.3 Strong Verbs

Use strong verbs instead of fancy-sounding ones. Examples:

- Verbose, weak verbs, are bad and short, strong ones, are good.
  - "Utilizes" change to "uses".
  - "Make an assumption" change to "assume".
  - "Is a function of" change to "depends on".
  - "Is an illustration" change to "illustrates, shows".
  - "Is a requirement" change to "requires, needs to".

5.2.4 Missing Articles

Check for missing articles:

- Roughly, concepts and classes of things do not use articles, most everything else more specific does. ("Routers transmit packets.")
- Do not use articles in front of proper nouns and names ("Firefox is a popular web browser.")
- Use articles before countable nouns ("devices", "cars")

5.2.5 Logical Connection of Sentences

Each sentence in a paragraph must have some logical connection to the previous one.

- For example, it may describe an exception ("but", "however"), describe a causality ("thus", "therefore", "because of this"), indicate two facets of an argument ("on the one hand", "on the other hand"), enumerate sub-cases ("first", "secondly") or indicate a temporal relationship ("then", "afterwards").

- If there are no such hints, check if your sentences are indeed part of the same thought. A new thought should get its own paragraph, but still clearly needs some logical connection to the paragraphs that preceded it.

5.3 Abbreviations and Acronyms

5.3.1 Abbreviations

Protocol abbreviations typically do not take an article, even if the expanded version does.

- For example, "The Transmission Control Protocol delivers a byte stream".
- But "TCP delivers a byte stream", since it is an abstract term.
- Observe that "The TCP design has been successful." is correct since the article refers to the design, not TCP.
- Note that abbreviation for organizations do take a definite article, as in "The IETF standardized TCP".
- Since the "P" in TCP, UDP and similar abbreviations already stands for "protocol", saying the "the TCP protocol" is redundant, albeit common. (LCD, Liquid Crystal Display, is another common case where many are tempted to incorrectly write LCD display.)

Check that abbreviations are always explained before use. Exceptions are tolerated, when addressed to the appropriate networking audience: ATM, BGP, ftp, HTTP, IP, IPv6, RSVP, TCP, UDP, RTP, RIP, OSPF, BGP, SS7.

- Sometimes, even basic terms, such as, PSTN and POTS are not taught to computer science students.
- For other audiences, even acronyms, such as, ATM should be expanded, as the reader might wonder why ATM has anything to do with cells, rather than bills.

5.3.2 Acronyms

When writing, expand all acronyms on first use, except acronyms that every reader is expected to know.

- In a research paper on TCP, expanding TCP is probably not needed
  - somebody who does not know what TCP stands for is not likely to appreciate the rest of the paper, anyway.

5.4 Hyphen

Use hyphens for concatenated words, such as, "end-to-end architecture", "real-time operating system."

- But "the computer may analyze the results in real time."
- Use "per-flow queuing", "flow-enabled", "back-to-back."

In general, hyphens are used in the following cases:

- To add prefixes that would result in double vowels (except for co-, de-, pre-, pro-), e.g., supra-auditory.
- All-: all-around, all-embracing.
- Half-: half-asleep, half-dollar (but halfhearted, halfway).
- Quasi-: quasi-public.
- Self-: self-conscious, self-seeking (but selfhood, selfless).
- To distinguish from a solid homograph, e.g., re-act vs. react, re-pose vs. repose, re-sign vs. resign, re-solve vs. resolve, re-lease vs. release.
- A compound adjective made up of an adjective and a noun in combination should usually be hyphenated. Examples: cold-storage vault, hot-air heating, short-term loan, real-time operating system, application-specific integrated circuit, Internet-based.
- Words ending in -like when the preceding word ends in 'l', e.g., shell-like.

5.4.1 Dashes

Do not overuse dashes for separation, as they interrupt the flow of words.

- Dashes may be appropriate to contrast robust thoughts, or the dash part is something not expected.
- The dash represents as long pause when speaking. In some cases, the used of a comma is better.
- If it is necessary to use a dash, make sure it is not a hyphen (- in LaTeX), but a real dash (— in LaTeX).

5.5 Miscellanea

Avoid the abuse of quotes, as they indicate that the writer is distancing himself from the term, or from the topic.

Numbers ten or less are spelled out: "It consists of three fields", not "3 fields." This is a universal standard, and a typical mistake in theses and articles.

Use until instead of the colloquial word till.

Use the following mathematical notation, with the proper capitalization: Equation (7), Table 10, Section 2, Chapter 3, Figure 23.

It is important to realize that optimal can not be improved – more optimally should be better or maybe more nearly optimal.

Avoid in-line enumeration like: "Packets can be (a) lost, (b) stolen, (c) delivered." The enumeration only interrupts the flow of thought.

It is not a good practice to refer to colors in graphs.

- Most people will print the paper on a monochrome (black and white) printer and will have no idea what the author is talking about.
- Make sure that graph lines are easily distinguishable, when printing on a monochrome printer. Use different line styles.

Do not forget to acknowledge the funding agency, the university or company that provided support to the research.

- If an author does not follow this recommendation, he may not have any to acknowledge in the future.

Check the references, to make sure they are up to date and complete.

- For example, conference papers might have been replaced by journal articles, and the subject of journal articles might have been included into books.

5.5.1 Bullets

Avoid itemization (bullets), as they take up extra space and make the paper paper look like a slide presentation, interfering with smooth reading.

- Bullets can be used effectively for emphasis of key points, such as, in this sentence.
- If one intends to describe components or algorithms, often the description environment works better, as it highlights the term, providing a low-level section delineation.
- Avoid long bulleted lists of one-sentence paragraphs.

5.5.2 References

Instead of "Reference [1] shows" or "[1] shows," use "Alencar [1] showed" or "Alencar and Tavares [1] showed" or "Alencar *et al* [1] showed" (if more than two authors). The Latin expression *et al*, that means "and others," is generally used for papers with more than two authors.

- Note that *et al* makes the subject plural, so it is "Smith *et al* [1] show" not "shows."
- Or, alternatively, "the Knuth protocol [1] is an example ...". This keeps the reader from having to flip back to the references, as they will recognize many citations by either author name or project name.
- No need to refer to RFC numbers in the text (except in RFCs and Internet Drafts). Exception is expected only for very low-level presentation: "RFC822-style addresses".

5.5.3 Capitalization

Use normal capitalization in captions, that should be treated as a regular sentence ("This is a caption," not "This is a Caption").

All headings must be capitalized consistently, either in heading style, capitalizing words, or sentence style, across all levels of headings.

- Generally, captions for figures and tables are best left in sentence style.

5.5.4 Parentheses and Brackets

Parentheses, brackets or braces are always surrounded by a space.

- Incorrect: The experiment(Figure 7)shows.
- Correct: The experiment (Figure 7) shows.

- Avoid excessive parenthesized remarks, because they make the text difficult to read.
  - Check whether the publication allows footnotes – some journals do not permit them.
  - More than two footnotes per page or several per paper is inconvenient for the reader.

5.5.5 Footnotes

The written material should make sense without the footnotes, that are dispensable at first reading.

- If the reader constantly has to look at footnotes, they are likely to lose their original place in the text.
- The best place for URL's is in the references rather than as a footnote, as the reader will know that the footnote is just a reference, not material important for understanding the text.

There is no space between the text and the superscript for the footnote. This is, in LATEX, the correct style is text[1], rather than text[2].

5.6 Syntactic Rules

It is important to follow a few syntactic rules, to improve the style and convey the right meaning in a technical text. For instance:

- "Respectively" is preceded by a comma, as in "The light bulbs, in the experiment, lasted 10 and 100 days, respectively."
- "Therefore" and "thus" are usually followed by a comma, as in "Therefore, the best idea should be implemented."
- Never use "related works" unless you are, for instance, talking about works of art. It is "related work".

Good writing style recommends that one should never start a sentence with "and".

- There are exceptions to this rule, but these are best left to English majors and novel writers.

Do not use colons (:) in mid-sentence.

- For example, "This is possible because: somebody said so" is wrong.
  - The part before the colon must be a complete sentence.

Do not start sentences with "That is because," and be careful not to confuse "its" with "it's" (it is).

In formal writing, contractions, such as, don't, doesn't, won't or it's are generally avoided.

5.6.1 Expressions

Vary expressions of comparison:

- "Flying is faster than driving" is better than "Flying has the advantage of being faster" or "The advantage of flying is that it is faster.".

---

[1]Correct footnote
[2]Incorrect footnote

Do not use slash-constructs such as "time/money," because they are imprecise, and can be confused with a division operation. This is acceptable for slides, but in formal prose such expressions should be expanded into "time or money" or "time and money", depending on the meaning intended.

5.6.2 Clichés and Symbols

It is difficult to accomplish, but try to avoid clichés like "Recent advances in ..." or "It is well-known that ..." to introduce a subject.

Do not use symbols, such as, "+" (for "and"), "%" (for "fraction" or "percentage") or "→" (for "follows" or "implies") in prose, outside of equations.

- These symbols are only acceptable in slides.

5.6.3 Paragraphs

Each paragraph should have a lead sentence summarizing its content. If this does not work naturally, the paragraph is probably too short.

- Try reading just the first lines of each paragraph – the paper should still make sense. For example, "There are only 10 types of people in the world: those who understand binary notation, and those who do not."

5.7 Units

Units are always in roman font, never italics or LaTeX math mode. Units are set off by one (thin) space from the number. In LaTeX, use ˜ to avoid splitting number and units across two lines. The commands \; or \, produce a thin space.

When dealing with mathematical notation, do not write ith, not $i - th$. For readability, powers of a 1,000 are divided by commas.

Use the units: shannon (Sh), that stands for "bit/s", or, if a metric prefix is used to indicate a multiple, use "kbit/s" or "Mbit/s" to represent transmission rates. Do not use "kbps" or "Mbps", because they are not scientific units, and can confuse the reader.

- Be careful to distinguish "Mbit" (Megabit) and "MB" (Megabytes), in particular "kbit" (1,000 bits) and "KB" (1,024 bytes).

As an example, the unit of frequency is always written as kHz (lower-case k), not KHz or KHZ. If in doubt, consult the Units and Measurements, Taligent Style Guide, or the NIST guide.

For time units, use "ms", not "msec", for milliseconds. Use "0.5" instead of ".5", i.e., do not omit the zero in front of the decimal point.

- It is recommended that, for quantities less than one, a zero should be set before the decimal point except for quantities that never exceed one.

5.8 Et Cetera

Avoid "etc.". Use "for example", "such as", "among others" or, better yet, try to give a complete list (unless citing, for example, a list of concepts known to be incomplete), even if abstract.

- *Et cetera* is not to be used for persons. It is equivalent to and the rest, and so forth, and hence not to be used if one of these would be insufficient, that is, if the reader would be left in doubt as to any important particulars.
- At the end of a list introduced by such as, for example, or any similar expression, the use of etc. is incorrect.

Thus, the correct way to express the idea is "software concepts like programs, codes and algorithms."

- The "like" and "for example" already indicate that there are more such items.
- Avoid excessive use of "i.e.". Vary the expression: "such as", "this means that", "because".
- Remember that the Latin expressions *id est* (*i.e.*) and *exempli gratia* (*e.g.*) are usually followed by a comma.
- Do not use ampersands (&) or slash-abbreviations (such as s/w or h/w) in formal writing. They are acceptable for slides only.

5.9 Figures and Tables

Use "in Figure 1" instead of "following figure" since figures may move, and even change pages, during the publication or typesetting process.

- Do not assume that the LATEXfigure stays where it is supposed to be.
- Text columns in tables are left-aligned, numeric columns are aligned on the decimal or right-aligned.

- Section, Figure and Table are capitalized, as in "As discussed in Section 3".
- Figure can be abbreviated as Fig., but Table is not usually abbreviated. This is a matter of taste, and it is important to be consistent
- Section titles are not followed by a period.

5.9.1 LaTeX Figures

In LaTeX use a tilde to tie the figure number to the reference, so that it does not get broken across two lines.

It is not advisable to use GIF images for figures, as GIFs produce low print quality and are huge.

- Export figures to PostScript.
- The Gimp, Xfig and Gnuplot programs generally produce PostScript figures that can be included without difficulties.

Only use line graphs when it is necessary to show a functional or causal relationship between variables.

- When showing different experiments, for example, use bar graphs or scatter plots.

5.9.2 Referring to Figures

Figures show, depict, indicate or illustrate something.

- Avoid writing, if possible, "(refer to Figure 17)."
- Often, it is enough to simply put the figure reference in parenthesis: "The linear system response (Figure 17) is obtained from the application of an impulse at its input."

5.9.3 Quotations and Citations

If a sentence from a reference is quoted literally, it is important to enclose it in quotation marks or show it indented and in smaller type ("block quote").

- A mere citation is not sufficient, as the reader does not know whether the author derived the material from the referred source or copied it *verbatim*.

Technical report citations must have the name of the organization, such as, the university or company, aside from other information that makes it possible for the reader to find the reference.

- Conferences must cite the location and date.

5.10 Who and Whom

The word "who" is a subjective, or nominative, pronoun, along with "he," "she," "it," "we," and "they." It is used when the pronoun acts as the subject of a clause. On the other hand, "whom" is an objective pronoun, along with "him," "her," "it," and "them."

It is used when the pronoun acts as the object of a clause. The use of "who" or "whom" depends on whether the author refers to the subject or to the object of a sentence. When in doubt, substitute "who" with the subjective pronouns "he" or "she." For example, "Who wrote the thesis?" can be answered as "He wrote the thesis."

Similarly, it is possible to substitute "whom" with the objective pronouns "him" or "her." For example, the sentence "I visited my supervisor whom I met in Canada," can be substituted by "I visited him."

5.11 Which and That

The use of "which" and "that" can lead to one of the most common mistakes in the language. The word "That" is a restrictive pronoun. It is essential to the noun to which it is referring. For example, "I do not trust a thesis that has no equations." In this case, I am referring to all theses. In other words, "I only trust theses that have equations."

The word "which" introduces a relative clause. It allows qualifiers that may not be essential. For example, "I recommend you study Mathematics, which is useful to write equations." In this case, you do not have to study a specific topic to learn how to write equations.

It is interesting to note that "which" qualifies, and "that" restricts. The term "which" is more ambiguous however, and by virtue of its meaning it is flexible enough to be used in many restrictive clauses. For instance, "The thesis, which you are reading, is mine." For example, "The thesis that received an award is mine.

5.12 Chapter Summary

This chapter deals with stylistic problems in English, ranging from grammar, to syntax, style, technical issues, expressions, figures, symbols and tables.

The reader can learn how to utilize these elements correctly and enhance readability, clarity and organization of the technical text. Problems with the correct use of verbs, abbreviations, acronyms, hyphen, units are also discussed. Some of the important points are:

- Talking and writing – Writing a technical text is somehow different from talking about the subject.
- Problems with verbs – Use consistent tense, prefer the present tense.
- Missing articles – Check for missing articles, verify the grammar rules that apply to the use of articles.
- Logical connection of sentences – Each sentence in a paragraph must have some logical connection to the previous one.
- Abbreviations – Protocol abbreviations typically do not take an article, even if the expanded version does.
- Acronyms – Expand all acronyms on first use, except acronyms that every reader is expected to know.
- Hyphen – Use hyphens for concatenated words: "end-to-end transmission, "real-time."
- Dashes – Do not overuse dashes for separation, as they interrupt the flow of words.
- Quotes – Avoid the abuse of quotes, as they indicate that the writer is distancing himself from the term, or from the topic.

5.13 Reference Material

- The Parallel Lives by Plutarch, published in volume V of the Loeb Classical Library edition, 1917.
    The text is in the public domain.
- *The Elements of Style*, by Strunk and White (Macmillan, 1979).
    - A good reference for a quick tutorial, that explains English prose writing in general.
- Handbook of Technical Writing, Gerald J. Alred, Charles T. Brusaw, Walter. E. Oliu. St. Martin's Press, New York, 2006.
    - Highly recommended; complete handbook on technical writing; entries arranged in alphabetical order; excellent index; English as a second language guidance; includes succinct guide to the writing process. Supporting material at `bcs.bedfordstmartins.com/-alredtech`.
- *The Encyclopedic Dictionary of Style and Usage*, by Mary A. DeVries. Berkley Books, 1999, New York, United States.

– A compact dictionary, with cross-references in place of a traditional subject index.

- *The Oxford English Dictionary of English Grammar*, by Sylvia Chalker and Edmund Weiner. Oxford University Press. 1994, London, United Kingdom.

  – This is a very useful reference book.

- *Modern Grammar*, by Paul Roberts. Harcourt, Brace & World, Inc. 1968, New York, United States.

  – This book presents the sentence-producing mechanism and examines the skeleton of the English sentence.

6

How to Prepare a Book

"If you want your children to be intelligent, read them fairy tales. If you want them to be more intelligent, read them more fairy tales."
Albert Einstein

The history of book writing goes back to the tablets used in Mesopotamia, in the 3rd millennium BC. The *calamus*, an instrument with a cross-section in the form of a triangle, was used to make characters in moist clay, which was then put in the fire, to give it some rigidity.

In the Ancient Egypt, the *papyrus* was used for writing, probably as early as from the First Dynasty, but the first evidence comes from the account books of King Neferirkare Kakai, of the Fifth Dynasty (ca. 2400 BC).

Printing began with the first bibles of Johannes Gutenberg, pressed in Mainz, German, in the late XV century. The first books used to replicate the handwritten books of the time, which varied widely. Over the centuries, publishers established conventions on the way books should be produced.

This chapter is a guide on the creation of a book. It presents a thorough list of items that comprise a general book. Few books would contain all the elements, but it is important to know they exist.

6.1 Parts of the Book

Books are generally divided into three main parts:

1. The front matter, contains the initial information about the book.
2. The body of the book, contains the essential information produced by the author.
3. The back matter, summarizes the book and gives other information.

Each part contains specific elements, and those elements should appear in a certain order. Probably, there is no book with all those parts, but the list is important to make sure the appropriate content is in the proper category, and that elements of the book appear in the sequence in which they are expected.

6.1.1 Front Matter

The pages at the beginning of a book, that appear before the body of the book. These pages are traditionally numbered with lowercase roman numerals.

- Half title – This page contains only the title of the book and is typically the first page seen when opening the cover. This page and its *verso* (the back, or left-hand reverse of the page) are often eliminated in an attempt to control the length of the finished book.
- Frontispiece – An illustration that is placed on the *verso*, facing the title page, or *recto*. The word comes from the Latin term *frontispicium*, meaning to look at the forehead.
- Title page – Announces the title, subtitle, author and publisher of the book. Other information that may be found on the title page can include the publisher's address, the year of publication, a description of the book, and illustrations.
- Copyright page – Usually the *verso* of the title page, this page carries the copyright notice, edition information, publication information, printing history, cataloging data, legal notices, and the book International Standard Book Number (ISBN).
  In addition, sequences of numbers are occasionally printed at the bottom of the page to indicate the year and number of the printing. Credits for design, production, editing and illustration are also commonly listed on the copyright page.
- Dedication – Some books carry a dedication, that follows the copyright page.
- Epigraph – An author may wish to include an epigraph, or quotation, near the front of the book. The epigraph may also appear facing the table of contents, or facing the first page of text. Epigraphs can also be used at the heads of each chapter.
- Table of Contents – Also known as the contents page, it lists all the major divisions of the book including parts, if the author uses this subdivision, and chapters. Depending on the length of the book, a greater level of detail may be provided to help the reader peruse the book.

The table of contents was invented by Quintus Valerius Soranus, before 82 BC.

- List of Figures – In books with numerous figures, or illustrations, it is necessary to include a list of all figures, their titles and the page numbers on which they occur.
- List of Tables – Similar to the mentioned list of figures, the list of tables is helpful for readers.
- Foreword – Traditionally a short piece written by someone invited by the author, the foreword puts the work in context. The foreword is always signed, and displays the person's name, place and date.
- Preface – Written by the author, the preface frequently tells how the book was created, and is often signed with the name, place and date.
- Acknowledgments – The author expresses their gratitude to those that helped in the production of the book.
- Introduction – The author explains the purposes and the objectives of the work, and may also place it in a context, as well as, describe the organization and scope of the book.
- Prologue – Is an opening to a story that establishes the context and gives background details. In a work of fiction, the prologue sets the scene for the story and is told in the voice of a character from the book, not the author's voice.
- Second Half Title – If the front matter is particularly extensive, a second half title identical to the first, can be added before the beginning of the text. The page following is usually blank but may contain an illustration or an epigraph. When the book design calls for double-page chapter opening spreads, the second half title can be used to force the chapter opening to a left-hand page.

6.1.2 Body Text

This is the text forming the main content of a book, thesis, magazine, journal, web page or other printed matter. A usual distinction between the parts of a book, is that the body matter is produced by the author, the front and back matter are made by the publisher.

- Part opening page – Both fiction and nonfiction books are often divided into parts, when there is a large conceptual, historical or structural logic that suggests these divisions.
- Chapter opening page – Most fiction and almost all nonfiction books are divided into chapters for the sake of organizing the material to be

covered. Chapter opening pages and part opening pages may be a single right-hand page, or in some cases a spread consisting of a *verso* and a *recto*.

- Epilogue – An ending piece, either in the voice of the author or as a continuation of the main narrative, intended to bring closure to the work.
- Afterword – A literary element that appears at the end of a piece of literature. Written by the author or by an invited person, it covers the origin of the book or situates the work in some wider context.
- Conclusion – A brief summary of the most noticeable arguments of the main work, to give a sense of completeness to the book.

6.1.3 Back Matter

At the end of the book various citations, notes and ancillary material are assembled into the back matter. It is, literally, all the material that follows the main text of a book.

- Postscript – From the Latin post *scriptum*, that means after the writing, relates to anything added to the main body of the work.
- Appendix or Addendum – A supplement to the main text. An appendix might include source documents cited in the text, material that arose too late to be included in the main body of the work, or a detailed mathematical deduction, to long to fit the corresponding chapter.
- Chronology – A chronological list of events, that may be useful for the reader. It can appear in the front matter if the author considers it critical to the understanding of the book.
- Notes – The final, or end notes usually come after any appendices, and before the bibliography or list of references. The notes are typically divided by chapter to make them easier to locate.
- Glossary – An alphabetical list of terms and their definitions, typically restricted to a specific area.
- Bibliography – A systematic list of books or other works, such as, articles in periodicals, usually cited in the main body of the document.
- List of Contributors – A work written by several authors may demand a list of contributors, which appears immediately before the index, although it is sometimes moved to the front matter. The names should be listed alphabetically by last name. Each contributor may have a brief biography, including academic affiliations and previous publications.

- Index – An alphabetical listing of people, places, events, concepts, and works cited along with page numbers indicating where they can be found within the main body of the work.
- Errata – A notice from the publisher of an error in the book, usually caused in the production process.
- Colophon – A brief notice at the end of a book usually describing the text typography, identifying the typeface by name along with a brief history. It may also credit the book's designer and other persons or companies involved in its physical production.

6.2 Writing a Book Proposal

Once the prospective author has an idea for a book, the next step is to begin to work on a proposal to be submitted to an editor or publisher. The purpose of a book proposal is to attract the editor's interest to publish the book.

The proposal provides a description, that could be used on the back cover of the book and in marketing materials. It must be written in a way that leaves a good impression, and must be an accurate account of the book contents.

It is important to guess why the publisher might want to publish this particular book, and think about the reason someone would be interested to read it. Some ingredients to include in any proposal are:

- Demand – Examine the published books that cover the same topics, ask what is missing and how this book will satisfy some specific public? Why should the publisher pick this particular book to open a new area?
- Market – Is the book directed for specialists in a certain field, or is it of interest to the general public? Can it be used as a textbook? If that is the case, indicate the course or discipline.
- Plan – Include a title and a table of contents, and estimate the size of each chapter, and how long it will take to complete the book. Also estimate the number of figures, tables, and other illustrations.
- Contributors – Define the book's authors or editors, and provide a short biography of each. Are these researchers attractive to the publisher, because they published well-known papers in good journals, or because they made significant discoveries?

Each publisher establishes specific requirements for book proposals. Therefore, it is essential to read the proposal guidelines before start writing. Book proposals are only a few pages long, but they require some work.

The time spent writing a good proposal saves time and helps produce a better book, because it improves the initial ideas for the book.

6.2.1 An Actual Book Proposal

In the following, it is presented a short version of the actual proposal for this book, submitted to the River Publishers editor Mark Jongh.

Proposal for a Book on Scientific Style in English to be Published by River Publishers

Title: Scientific Style in English

Authors Names and Affiliations

Marcelo Sampaio de Alencar, Thiago Tavares de Alencar
Institute for Advanced Studies in Communications (Iecom)
Federal University of Campina Grande
Av. Aprigio Veloso, 882, Bairro Universitário
58.429-900 Campina Grande PB Brazil
E-mail: malencar@dee.ufcg.edu.br
Phone: +55 83 21011410 (Office)
Fax: +55 83 21011418

Background

Scientific style is a new topic in the educational market, which evolved from the amalgamation of different areas.

In commercial terms, publication is a driving force of the economy and its influence throughout the world will generate a huge market, which includes the media industry, the editing companies, the universities and colleges.

Brief Description of the Book

Good scientific writing must be scientifically accurate, concise, useful, clear, illustrated with visuals, targeted to a specific audience, well organized, interesting, consistent, complete, correct in spelling, punctuation and grammar. The book presents information and tips on how to read, write and make oral presentations in a scientific style.

Reasons for Writing the Book

It is important to realize that the academy, the media industry, the information technology sector and the companies, which deal with enormous sums of money each year, rely on badly written papers and technical reports. The employment market in this area is in demand of professionals of information technology and engineers who can learn new techniques, but also write standards, articles and theses.

The books that have been published so far do not cover all the subjects needed to independently produce quality papers and reports.

Market and Readership

This book is aimed at senior undergraduate students, graduate students, engineers and information technology professionals in the areas of Electrical Engineering and Computer Science. It can be used as a textbook for a course on Scientific Writing. No background is expected from the student.

Table of Contents

This section describes the book body, including all the chapters, with sections, the appendices, the bibliography and remissive index.

Competition

The following list presents the competing books on the market at the time this proposal was written. Most of the books in the list are cited as references in the proposed book Scientific Style in English, but none cover all the topics, ranging from scientific reading and writing, to presenting. Follows a long list of references.

- Periodicals and handbooks,
- Reference books,
- Popular words and spelling,
- Dictionaries,
- The TEX and LATEX reference books,
- Grammars,
- Guides to writing,
- Online writing guides,
- Writing references.

Manuscript Information

The book will have approximately 150 printed pages, and the final manuscript could be delivered in eighteen months. It has no figures, as of now.

Author's Information

This section contains the short biographies of the authors, which include current affiliation, education, books, chapters and paper published, awards received, societies they belong, and other important information to captivate the potential readers.

6.3 Chapter Summary

This chapter teaches the basic steps for preparing a book. It displays all the elements that can possibly exist in a book, and focuses on the main parts which all books usually have.

Books are generally divided into three main parts, the front matter, the body and the back matter, described in the following:

1. The front matter – comprises the pages at the beginning of a book.

   (a) Half title – The first page, that contains only the title of the book.

   (b) Frontispiece – An illustration placed on the *verso*.

   (c) Title page – Contains the title, subtitle, author, publisher's address, year of publication, a description of the book, and illustrations.

   (d) Copyright page – Carries the copyright notice, edition information, publication information, printing history, cataloging data, legal notices, and the book ISBN.

   (e) Dedication – The dedication follows the copyright page.

   (f) Epigraph – May also appear facing the table of contents.

   (g) Table of Contents – Lists all the major divisions of the book including parts, chapters and sections.

   (h) List of Figures – This list includes titles and the page numbers on which the figures occur.

   (i) List of Tables – Similar to the list of figures.

   (j) Foreword – A short piece that puts the work in context.

(k) Preface – It preface tells how the book was created, and gives additional information.

(l) Acknowledgments – Expresses the author gratitude to those that helped in the production of the book.

(m) Introduction – Explanation of the purposes and the objectives of the work.

(n) Prologue – An opening to a story that establishes the context and gives background details.

(o) Second Half Title – A second half title identical to the first, can be added before the beginning of the text.

2. The body of the book – The main parts of the book.

(a) Part opening page – Fiction and nonfiction books can be divided into parts.

(b) Chapter opening page – Books are usually divided into chapters, to organize the material.

(c) Epilogue – An ending piece, intended to bring closure to the work.

(d) Afterword – An element that appears at the end of a piece of literature.

(e) Conclusion – A brief summary of the most noticeable arguments of the main work.

3. The back matter – The final parts of the book.

(a) Postscript – Relates to anything added to the main body of the work.

(b) Appendix – A supplement to the main text.

(c) Chronology – A chronological list of events.

(d) Notes – End notes that come after the appendices.

(e) Glossary – An alphabetical list of terms and their definitions.

(f) Bibliography – A systematic list of books or other works.

(g) List of Contributors – Each contributor may include a short biography.

(h) Index – An alphabetical listing of people, places, events, concepts, and works cited along with page numbers in which they appear in the text.

(i) Errata – A notice from the publisher of an error in the book.

(j) Colophon – A brief notice at the end of a book that describes the text typography,

6.4 Reference Material

- *The Elements of Style*, by Strunk and White (Macmillan, 1979).
    - An excellent reference for a quick tutorial. The prospective authors should read this book, which explains English prose writing.
- *A Handbook for Scholars* by Mary-Claire van Leunen (Knopf, 1978).
    - This well-written book is a pleasure to read. It explains the use of footnotes, references and quotations.
- Handbook of Technical Writing, Gerald J. Alred, Charles T. Brusaw, Walter. E. Oliu. St. Martin's Press, New York, 2006.
    - It is highly recommended. A complete handbook on technical writing, with entries arranged in alphabetical order, an excellent index. Includes English as a second language guidance and a succinct guide to the writing process. The supporting material can be found at bcs.bedfordstmartins.com/alredtech.
- The TEXbook (Computers & Typesetting), by Donald E. Knuth. Addison-Wesley Professional. 1984, New York, United States.
    - This is the reference book on TEX written by the master himself.
- LATEX A Document Preparation System (2nd Edition), by Leslie Lamport Addison-Wesley Series on Tools and Techniques for Computer T, 1994, New York, United States.
    - This is the reference book on LATEX and has many examples.

7

Theses and Dissertations

"If we knew what it was we were doing, it would not be called research, would it?"
Albert Einstein

This chapter covers the subject of theses and dissertations, similar terms that represent the required submissions for a master's degree and doctorate. Different universities use the words in distinct manners, some call the doctoral formal treatise requirement a thesis, others call it a dissertation. And the same is true for the master's written report.

This chapter includes guidelines and general recommendations for writing a thesis or dissertation, which are useful for students of any course.

7.1 Thesis Definition

A thesis is a research report, formally approved by a university. It is a monograph, a self-contained piece of work, written solely by the candidate, who has to declare it.

The word came from the Greek term for proposition. The report concerns a problem or a set of problems in a certain area of research, and it should describe what was known about it previously, what the author did to solve it, what the results mean, and how further progress in the field can be made.

Most universities award a Ph.D. degree if the candidate demonstrates ability to develop independent research that attain academic standards. Typically, a successful Ph.D. candidate shows this ability by the study of a particular area within a field for three to five years, and having made at least one new discovery, or at least one contribution, to the knowledge of a sub-area within the chosen area of the subject

A thesis is also supposed to be used as a scientific report, and will be consulted by future workers in the field, or consulted by people from other institutions. Theses are now stored in digital form in government databases and university libraries, to be available to the general public.

In the technological arena, a thesis usually involves a mathematical, logical or computational model of a certain phenomenon, or a solution to a complex problem. In general, it is an academic response to the previously posed hypothesis.

7.2 Thesis Description

The text of a thesis must be clear, and present good grammar and logical writing, to make it easier to be read by students and scholars. Scientific writing has to be formal, as discussed in previous chapters, but also elegant and efficient in communicating the idea.

To obtain a doctorate, it is necessary to write a thesis about an interesting theme, put some independent and novel contributions in context, and compare them critically with other approaches.

The candidate must also defend the thesis, It may be done in a viva, which is a discussion with examiners, who are experts in the area, or in a public presentation, that usually includes internal and external examiners.

A basic description of a thesis, or dissertation, generally includes:

- Abstract – A short account of the main results of the thesis. Sometimes written in two languages.
- Introduction – Presents the structure of the thesis, state the contributions, to place and define the problem. It has to motivate the problem, propose the terminology, describe acronyms, and state the hypothesis, sometimes using concrete examples and figures, quoting data sources, that is, industry analysts, market surveys or case studies.

  It is important to provide a list of the contributions, but do not use hyperbole, such as, adjective or intensity adverbs, when referring to the work, for example, avoid: excellent results, highly reliable, magnificent conclusions, extremely efficient. Never try to sound like the thesis covers an entire field or domain.
- Related Work – Presents a survey and critical assessment of the papers, reports or books related to the main subject of the dissertation. List the key papers, including the seminal and the recent work. and acknowledge the role that each paper has played in the field.

  The critic of each article has to be constructive, only to the extent that it justifies the approach taken, but never diminish a paper, just to show some knowledge in the area, and do not make the criticism personal.

- Analysis, Modeling and Results – It is essential to be clear and impartial about the assumptions and limitations of the research, and state the requirements and constraints of the proposed approach. Establish a validation plan for the hypothesis, which may include a set of experiments, use of simulation, demonstration of theorems and proofs, and define the scope of the thesis.

  Do not expect to write about all of the work that was done during the graduate research career, do not present a set of scattered, unrelated results just to fill space, do not trumpet all of the advantages of the approach, in a repetitive way, and do not make promises that are not supported by the results.

- Conclusions – It is essential to be reflective and honest about the research, state the lessons learned, point the the overall insights, discuss the completion of the work, the progress made in the field because of the thesis.

- Further Work – Presents a list of subjects to be explored by other researchers, and do not consider this chapter as a list of the limitations of the thesis.

- Appendix – Includes extra material, lengthy deductions, programming code, case studies.

- Bibliography – Must be complete and include all the information that a reader eventually will need to find the papers, reports and books used in the thesis preparation.

7.3 The Thesis Structure

The following structure is appropriate for some theses or dissertations, but some of the items may be unrelated to a particular area. Sometimes, results and discussion are presented in separate chapters, or in the same chapter.

It is useful to make a plan of each chapter and section before starting to write, because it makes the thesis easier to write, and permits writing the chapters in parallel, instead of writing them in sequence.

By the way, this is an important point that worth to be reinforced. Although the final version of a thesis has the chapters placed in a logical sequence, they are usually written in a different, sometimes, random, order.

For instance, the introductory part is commonly written at the very end of the research work, when the candidate has a good understanding of the problem, and of the possible solutions. It is the same with the abstract and the list of symbols, that come at the beginning of the manuscript.

Therefore, do not wait to finish a chapter to start writing the next one! Write them in parallel, as soon as the ideas come to the mind.

7.3.1 Copyright

Every institution has to comply to a certain level of bureaucracy. This standard page gives the university library the right to publish the work, and make it available to other researchers, possibly by microfilm, in electronic format, or by other medium.

7.3.2 Declaration

It is required by the institution to guarantee that the thesis was written by the author alone. Universities request a statement such as: "I hereby declare that this submission is my own work and that, to the best of my knowledge and belief, it contains no material previously published or written by another person, nor material which, to a substantial extent, has been accepted for the award of any other degree or diploma of the university or other institute of higher learning, except for source material appropriately acknowledged."

7.3.3 Title Page

The title page may vary for different institutions, but it must include the title, the author, the date, and a statement, such as: "A thesis submitted for the degree of Doctor of Philosophy in the Polytechnic School, Federal University of Bahia."

7.3.4 Abstract

This part is the most widely published and most read, because it is sent to government databases, and published in Dissertation Abstracts International. It should be written when the work is done. It is a succinct description of the thesis, including the methodology, main results and conclusions.

7.3.5 Acknowledgments

Authors usually write a page of thanks to those who helped with the preparation of the thesis, with the research, with the experiments or with something that is deemed important. It is not uncommon to acknowledge the help of family, friends, university colleagues and staff, and, of course, the supervisor.

7.3.6 Table of Contents

The table of contents usually lists the chapters, including sections and subsections, the appendices, the remissive index, if any, and the bibliography. The first pages of the thesis are in roman numerals.

7.3.7 Introduction

This chapter describes the topic and its importance. It is written for researchers in the general area, which are not, normally, specialists in the subject of the thesis. The introduction must hold the attention of the reader.

7.3.8 Literature Review

The candidate should start collecting research material as soon as possible, preparing summaries of the papers, and storing reference information in an appropriate form, for instance, in a BibTeX file, for Linux lovers, or in a spread sheet, for Windows users.

The author must demonstrate what is currently known about the problem, and which other methods have been tried to solve it. Do not omit relevant papers, authored by researchers who are likely to be in the thesis committee, or seminal articles in the area. On the order of a hundred papers is a reasonable number of references, but it depends on the field.

7.3.9 Middle Chapters

All of the interesting and relevant data are placed in the middle chapters. The experiments are described in detail in a thesis, because it is an axiom of science that any result must be reproducible. It is common for a researcher to request a thesis to obtain more detail on how a study was performed, to reproduce it. A thesis is expected to have far more detail than a journal article.

The exact structure of the middle chapters differ, depending on the subject. In some theses, it is necessary to build a theory, to describe the experimental techniques, then to report what was done on diverse problems, and then to present a model or a theory based on the new findings.

For this type of thesis, the chapter title may be related to the theory, the materials and the methods, the proposed theory or model. A conclusion chapter is also mandatory. In other cases, it would be more appropriate to discuss different techniques in different chapters, rather than to have a single chapter that associates them.

In the following there are some comments on the elements materials and methods, theory, results and discussion, which may or may not correspond to thesis chapters.

Within each chapter it is important to include some essential parts, to introduce the topic do be discussed, to approach the problem and to correlate with the next chapter.

- Introduction – Defines the objectives of the chapter, state the problem or issue it addresses, and describe how the chapter fits within the overall script of the thesis.
- The Content – Establish a rigorous approach to the problem, or present a detailed explanation of the issue, pointing out the assumptions underlying the problem, or indicating a complete description of the issue. Delineate the validation scheme, either by system design, new theory, implementation, use of graphs or references.
- Summary – Repeat the important points of the chapter, include a transition sentence to introduce the next chapter.

Materials and Methods

This chapter may be absent in theoretical theses, but the important point it that a researcher should be able to reproduce exactly what has been done, following the description.

Theory

A theoretical work must include sufficient material to allow the reader to understand the arguments and their physical bases, but, it is not necessary to reproduce material that the reader could find in a standard text. Do not include theory that is not related to the research.

Physical arguments are as important as equations, which must be explained, term by term. The important, special, cases must be emphasized.

A proper theoretical work may include a certain amount of detail, but consider moving lengthy derivations to appendices. The order and style of presentation are important, and the order in which the work was done may not be the best presentation order.

Results and Discussion

Results and discussion are usually combined in theses. The conditions and constraints that permitted to obtain the results must be established in the beginning, as well as, all the relevant parameters. Use appropriate statistical analyses and tests, present measurement errors on the graphs or tables.

The origin and intercepts of a graph are important, the zeros of one or both scales should appear on the graph. Each coordinate must have a name and a unit. The curves should be drawn using different styles for the lines, and have associate legends. The caption must describe the illustration.

Results usually require discussion, and the author should comment on their meaning, on how they fit into the area of knowledge, and on their consistency, regarding current theories. It is important to tell if the results give new insights into the research theme, or suggest new theories or methods.

7.3.10 Final Chapter, References and Appendices

This part of the thesis includes the final chapter, that presents and discusses the conclusions, the complete list of references and the optional appendices.

Conclusions and Suggestions for Further Work

A summary of conclusions is usually longer than the final part of the abstract, and the author can be more explicit and more detailed about the work. Scientific investigation, sometimes, raise more questions than produce answers.

To write this chapter the author should inquire if the work suggests any interesting further branches of knowledge to pursue, if there are ways in which the research could be improved by future researches, and what are the practical implications of the work.

References

Reference all the cited journal articles, thesis, reports and books, but also web sites, mentioning the data of the access and giving its Universal Resource Locator (URL).

Appendices

The appendix includes material that should be in the thesis but which would disrupt the information flow. Things which are typically included in appendices are: original computer programs, data files, mathematical demonstrations, pictures or diagrams that do not fit the main text.

7.4 Chapter Summary

This chapter covers the subject of theses and dissertations, that are required submissions for a master's degree and doctorate.

The chapter includes guidelines and general recommendations for writing a thesis or dissertation, which are useful for students of any course.

A basic description of a thesis, or dissertation, generally includes:

- Abstract – A short account of the main results of the thesis.
- Introduction – Presents the structure of the thesis, the contributions, and the problem.
- Related Work – Presents a survey and critical assessment of the literature.
- Analysis, Modeling and Results – States the requirements and constraints of the proposed approach, and establishes a validation plan for the hypothesis.
- Conclusions – States the lessons learned, the the overall insights, discusses the work, and the progress made in the field.
- Further Work – Presents a list of subjects to be explored by other researchers.
- Appendix – Includes extra material, lengthy deductions, programming code, case studies.
- Bibliography – Includes all the information that a reader needs, to find the papers, reports and books used in the thesis preparation.

7.5 Reference Material

- How to Write a Good Ph.D. Thesis and Survive the Viva, by Stefan Rüger. Knowledge Media Institute, The Open University, Buckingham, United Kingdom, 2016.
- Doing postgraduate Research in Australia, by K. Stevens and C. Asmar. Melbourne University Press, Melbourne, Australia, 1999.

- How to Get a PhD: A Handbook for Students and their Supervisors, by E. M. Phillips and D. S. Pugh. Open University Press, Buckingham, England, 1994.
- The Visual Display of Quantitative Information, by E. R. Tufte. Graphics Press LLC, Cheshire, United States, 1983.
- Envisioning Information, by E. R. Tufte. Graphics Press LLC, Cheshire, United States, 1990.
- *Mathematical Writing*, by Donald E. Knuth, Tracy Larrabee, Paul M. Roberts. Mathematical Association of America, 1989.

  – This book is based on a course given by Donald Knuth at Stanford University. It discusses technical writing and presentation of mathematics and computer science, including preparation of theses, papers, and books.

- An Outline of Scientific Writing: For Researchers With English as a Foreign Language, J. T. Yang. World Scientific, 1995.

  – May be especially useful to English as a second language writers.

- MIT Guide to Science and Engineering Communication, J. G. Paradis and M. L. Zimmerman. MIT, Cambridge, 2002.

  – Discusses all types of technical communication and includes a list of 27 guidelines for style and usage.

- The TEXbook (Computers & Typesetting), by Donald E. Knuth. Addison-Wesley Professional. 1984, New York, United States.

  – This is the reference book on TEX and every serious author must have it.

- LATEX A Document Preparation System (2nd Edition), by Leslie Lamport Addison-Wesley Series on Tools and Techniques for Computer T, 1994, New York, United States.

  – This is the reference book on LATEX and every prospective author must have it.

8

Suggestions for Public Speaking

"If you can't explain it to a six year old, you don't understand it yourself."
Albert Einstein

It is important to know that oral communication is different from written communication, because the audience has only one chance to hear the talk. Therefore, make an effort to be clear and pass on the right message.

It is not uncommon for the listeners to attend several talks on the same day, particularly if they attend a conference or meeting. It is important to state the points clearly, mainly if the audience is not allowed to ask questions during the presentation.

Cognition experts indicate that it is easier to remember something if the key points are repeated, at least three times, in different ways.

To think about the audience, to acquire some empathy, is of paramount importance, because some audiences must be addressed regarding distinct levels of knowledge. Some are experts in the subject, some are experts in the general area, and others know little or nothing about the subject.

Therefore, explain subject to the experts, but make the introduction and the summary accessible to the whole audience. This chapter presents a few tips on presentation for public speaking.

8.1 Introduction

The needs of the audience are important to accommodate the contents of the presentation. The material must be known thoroughly, and has to be arranged in a logical sequence.

Make sure the speech is captivating to the audience, as well as worth their time and attention. Practice and rehearse the speech, in front of your family, friends or colleagues. Videotape the presentation, if possible, watch and analyze it. Emphasize personal strong points during the presentation.

8.2 In Front of an Audience

When presenting in front of an audience, the speaker performs as an actor on stage, therefore, how the public perceives the acting is decisive. It is important to dress appropriately for the occasion and be solemn if the topic is serious.

A good speaker presents the proper image to the audience, he or she should look pleasant, enthusiastic, confident and proud to talk about the subject. It is advisable to remain calm, appear relaxed, even if feeling anxious.

The initial pace of the speak is slow, to enunciate clearly the topics, but it is essential to show appropriate emotion and feeling relating to each topic. Establish a certain empathy with the audience.

8.3 How to Adjust the Tone

It is crucial to speak to the person at the back of the auditorium, to ensure the voice tone is loud enough to be heard by everyone. An experienced speaker varies the voice tone and dramatize, if necessary, to convey the right information. Of course, if a microphone is available, run a previous test, and adjust the sound system accordingly.

8.4 Body Language

Every actor knows that body language is meaningful. It is better to stand, walk and move about with corresponding hand gesture or facial expression. It is not advisable to sit down or stand still, just reading from a prepared speech.

A speaker must show conviction and believe in the subject being presented. It is vital to persuade the audience effectively. The material to be presented orally and the written research paper should include the same ingredients.

The presentation progresses from the introduction, in which the thesis is stated, to the main text, that introduces strong supporting arguments, accurate and up-to-date information, to the conclusion, in which the thesis is re-stated, a summary is presented, and the logical conclusion is finally drawn.

8.5 Presentation Equipment

The use of audio-visuals, including slides or videos, is recommended. It is also significant, if appropriate, to use the blackboard, at times, to explain a topic in detail. This is important when presenting a talk in front of a selection committee.

A speaker should master the use of presentation software, such as, Beamer, in case of a Linux user, or PowerPoint, if a Windows user, to do the presentation. It is not recommended to fill the audience with excessive use of animation, sound clips, or the whole pallet of colors, if not appropriate and necessary for the topic.

The speaker must know the adequate size of the fonts, considering the room dimension, and not torment the audience by presenting a lengthy document in tiny print on an overhead that has to be read out loud.

8.6 Presentation Tips

The following points indicate some interesting tips, to aid the presentation.

- It is important to not keep reading from notes for a long time, although it is acceptable to glance at the notes once in a while.
- If possible, make a positive comment about the results of the previous speaker, if they are related to the subject. Do not make negative statements about any of the speakers.
- It is essential to speak loudly and clearly, sound confident, show good humor, and do not murmur.
- If an error is committed, just correct it, and continue. There is no need to make excuses or apologize.
- Maintain a genuine eye contact with the audience. It it influential to look straight into the eyes of a person in the audience for a few seconds at a time. Make direct eye contact with a number of people in the audience, and frequently glance at the whole audience while speaking, to make everyone feel involved.
- Speak to the audience, listen to the questions, respond to the reactions, adjust the tone to the response of the public, and change the strategy if necessary, to keep the attention of the spectators.
- Remember that communication is critical to a successful presentation.
- If there is a shortage of time, it is relevant to know what can be safely ignored.
- If there is some time left, decide what could be effectively added.
- Always be prepared for the unexpected, and keep a few extra slides, in case something happens.
- Pause, and allow the audience to have some time to reflect about the subject. Do not speed up the presentation if the audience is feeling out of breath.

- Sometimes, it is important to introduce a bit of humor, whenever appropriate and possible, but do not tell jokes, if they are not related to the subject.
- Keep the audience interested throughout the entire presentation, because an interesting speech makes time fly, but a boring speech is always long to endure.
- Whenever possible, it is advisable to give the audience handouts at the appropriate time. Tell audience ahead of time they will receive an outline of the presentation, so that they do not need to take unnecessary notes during the presentation.

8.7 Final Tips

As final comments on presentation, when using audio-visuals, be sure all necessary equipment is set up and in working condition, prior to the presentation. If possible, provide an emergency backup system that is readily available.

Check out the location *a priori* to ensure seating arrangements for the audience, Verify if the flip-chart, the whiteboard, the lighting, the location of the projection screen, and the sound system are suitable for the presentation.

It is imperative to know when to finish the talk. Use a timer, a clock, or a smartphone, to time the presentation when preparing it.

Do not bore the audience with repetitious or unnecessary words in the presentation. At the end of the presentation, summarize the main points in the same way as is usually done for a written paper.

Conclude the presentation with an interesting remark, or an appropriate punch line, to leave the listeners with a positive impression and a sense of fulfillment. Thank the audience, ask for questions, answer them, wait for applause, and leave the stage.

8.8 Chapter Summary

This chapter covers the topic of public speaking, with many useful tips, such as explaining the subject matter to an audience of mixed knowledge levels on the subject, to be interesting for experts and beginners alike.

It also shows how to use body language and tone to enhance the presentation, how to best maintain eye contact and preparing and utilizing the audiovisual and multimedia materials for the presentation.

8.9 Reference Material

- *IEEE Transactions on Professional Communication.*
  - Published since 1958 to improve professional communication among the IEEE members.
- MIT Guide to Science and Engineering Communication, J. G. Paradis and M. L. Zimmerman. MIT, Cambridge, 2002.
  - Discusses all types of technical communication and includes a list of 27 guidelines for style and usage.
- Basic Public Speaking, 2nd Edition (Paperback): The Roadmap to Confident Communications, by Douglas A. Parker. Xlibris Corporation, Bloomington, United States, 2001.
  - The book offer the reader an informative, accessible, succinct, and practical manual to acquire skills, expertise and confidence in addressing the public, from informal groups to formal conference meetings.

Recommended Online Guide

- A Research Guide for Students: Presentation Tips for Public Speaking, by I. Lee, Sheli Rodney, Alexander Rodney, Simon Rodney. 2014. `http://www.aresearchguide.com/3tips.html`.

A

Assignments

"Most teachers waste their time by asking questions that are intended to discover what a pupil does not know, whereas the true art of questioning is to discover what the pupil does know or is capable of knowing."
Albert Einstein

A.1 Introduction

It is important to mention that style is a distinctive manner of expression, in writing or speech, the way in which something is said or done, as distinguished from its substance, and is also a convention with respect to spelling, punctuation, capitalization, and typographic arrangement and display followed in writing or printing.

This appendix presents a few assignments to review the text, and stimulate the reader to start practicing his own style.

A.2 Assignments

- Assignment 1
    - Plan the writing of a one-page article to be submitted to an international conference.
- Assignment 2
    - Write the *Abstract* of an article to be submitted to a journal.
- Assignment 3
    - Write the *Introduction* of an article to be submitted to a journal. Include *References*.

- Assignment 4
    - Write an article to be submitted to a journal. Include all sections.
- Assignment 5
    - Outline a *Status Report* of your own research.
- Assignment 6
    - Write a *Status Report* of your own research. Include all important sections.
- Assignment 7
    - Write a *Book Proposal* to an Editor.
- Assignment 8
    - Review an *Article*, which is full of errors, and score for each corrected error.

Bibliography

[1] Abramowitz, M., and Stegu, I. A. (Ed.). (1965). *Handbook of Mathematical Functions*. New York, NY: Dover Publications Inc.,

[2] Adler, M., and van Doren, C. (1972). *How to Read a Book*. New York, NY: MJF Books.

[3] Alred, G. J., Brusaw, C. T., and Oliu, W. E. (1995). *Handbook of Technical Writing*. New York, NY: St. Martin's.

[4] American Heritage Publishing Company (1969). *American Heritage Dictionary of the English Language*. Boston, MA: Houghton Mifflin Harcourt.

[5] Babington, D., and LePan, D. (2002). *Broadview Pocket Guide to Writing*. Toronto, ON: Broadview Press.

[6] Baron, N. S. (2000). *Alphabet to Email: How Written English Evolved and Where It's Heading*. London: Routledge.

[7] Blake, G., and Bly, R. W. (1993). *The Elements of Technical Writing*. London: Macmillan.

[8] Boyer, C. B., and Merzbach, U. C. (2011). *A History of Mathematics*, 3rd Edn. Hoboken, NJ: John Wiley & Sons, Inc.

[9] Chalker, S., and Weiner, E. (1994). *The Oxford English Dictionary of English Grammar*. London: Oxford University Press.

[10] Day, R. A. (1979). *How to Write and Publish a Scientific Paper*. Cambridge: Cambridge University Press.

[11] DeVries, M. A. (1999). *The Encyclopedic Dictionary of Style and Usage*. New York, NY: Berkley Books.

[12] Dunham, W. (1991). *Journey through Genius – The Great Theorems of Mathematics*. London: Penguin Books Ltd.

[13] Dunham, W. (2005). *The Calculus Gallery – Masterpieces from Newton to Lebesgue*. Princeton, NJ: Princeton University Press.

[14] Sparks, A. N., (Senior Ed.). (1992). *Webster's New Word – Speller/Divider*. Webster's New World, 2 Edn, New York, NY.

[15] Shallit, J. O. (Ed.) (2017). *LaTeX Style Guide for the Journal of Integer Sequences.* Waterloo, ON: Journal of Integer Sequences.

[16] Broquist, L. (Senior Ed.) (2017). *Wordsmyth Dictionary and Thesaurus.* Ithaca, NY: The Wordsmyth Organization.

[17] Hazewinkel, M. (Ed.) (2002). *Encyclopaedia of Mathematics.* Dordrecht: Kluwer Academic Publishers.

[18] Forman, J., and Kelly, K. (1990). *The Random House Guide to Business Writing.* New York, NY: McGraw-Hill.

[19] Fowler, H. W. (1908). *The King's English.* Oxford: Clarendon Press.

[20] Gillman, L. (1987). *Writing Mathematics Well: A Manual for Authors.* Washington, DC: Mathematical Association of America.

[21] Halmos, P. R. (1973). *How to Write Mathematics.* Providence, RI: American Mathematical Society.

[22] Houp, K. W., and Pearsall, T. E. (1992). *Reporting Technical Information.* New York, NY: MacMillan Publishing Company.

[23] Strunk, W. Jr. (1918). *The Elements of Style.* Ithaca. NY: Harcourt, Brace & Howe.

[24] Strunk, W. Jr., and White, E. B. (1979). *The Elements of Style.* New York, NY: Macmillan Publishers.

[25] Knuth, D. E. (1984). *The TpXBook (Computers & Typesetting).* Reading, MA: Addison-Wesley Professional.

[26] Knuth, D. E., Larrabee, T., and Roberts, P. M. (1989). *Mathematical Writing.* Washington, DC: The Mathematical Association of America.

[27] Lamport, L. (1994). *LaTeX A Document Preparation System,* 2nd Edn. Reading, MA: Addison-Wesley Series on Tools and Techniques for Computer.

[28] Lindsell, S. L. (1989). *The Secretary's Quick Reference Handbook.* Brooklyn, NY: Arco.

[29] Longman (2012). *The Longman Dictionary of Contemporary English.* London: Pearson.

[30] MacAllister, J., and Madama, G. (1976). *English for Electrical Engineers.* London: Longman.

[31] Mccaskill, M. K. (1990). *Grammar, Punctuation, and Capitalization: A Handbook for Technical Writers and Editors.* Washington, DC: NASA Langley Research Center.

[32] Merriam-Webster (1985). *Webster's Standard American Style Manual.* Springfield, MA: Merrian-Webster Inc.

[33] Merriam-Webster (1994). *Merriam-Webster's Dictionary of English Usage.* Springfield, MA: Merriam Webster.

[34] Merriam-Webster (2014). *Merriam-Webster Dictionary and Thesaurus.* Springfield, MA: Merriam-Webster.

[35] Oxford (2017). *The Oxford English Dictionary.* Oxford: Oxford University Press.

[36] Paradis, J. G., and Zimmerman, M. L. (2002). *MIT Guide to Science and Engineering Communication.* Cambridge, MA: Massachusetts Institute of Technology.

[37] Parker, D. A. (2001). *Basic Public Speaking: The Roadmap to Confident Communications*, 2nd Edn. Bloomington, IN: Xlibris Corporation.

[38] Parkes, M. B. (1992). *Pause and Effect.* Aldershot: Scholar's Press.

[39] Penrose, J. M., Rasberry, R. W., and Myers, R. J. (1993). *Advanced Business Communication.* Balmont, CA: Wadsworth Publishing Company.

[40] Perelman, L. C., Barrett, E., and Paradis, J. (2001). *Mayfield Handbook of Technical and Scientific Writing.* New York, NY: McGraw-Hill Higher Education.

[41] Phillips, E. M., and Pugh, D. S. (1994). *How to Get a PhD: A Handbook for Students and their Supervisors.* Buckingham: Open University Press.

[42] Plutarch (1917). *The Parallel Lives,* Trans. Bernadotte Perrin. Loeb Classical Library edition, Vol. 5, Cambridge, MA: W. Heinemann.

[43] Dundurn Press (1997). *The Canadian Style – A Guide to Writing and Editing.* Toronto, ON: Dundurn Press.

[44] Purdue University (2016). *The Purdue Online Writing Lab (OWL).* West Lafayette, IN: Purdue University.

[45] Beyer, R. T. (1990). *AIP Style Manual.* Woodbury, NY: American Institute of Physics.

[46] Roberts, P. (1968). *Modern Grammar.* New York, NY: Harcourt.

[47] Roget, P. M. (1852). *Roget's Thesaurus.* Essex: Longman Group Limited.

[48] Ruger, S. (2016). *How to Write a Good Ph.D. Thesis, and Survive the Viva Class Notes.* London: The Open University.

[49] Lester, J. D. (1984). *Writing Research Papers – A Complete Guide.* Glenview, IL: Scott Foresman and *Company.*

[50] Siegel, J. G., and Shim, J. K. (1987). *Dictionary of Accounting Terms.* Hauppauge, New York, NY: Barron's Educational Series Inc.

[51] Stevens, K., and Asmar, C. (1999). *Doing Postgraduate Research in Australia.* Melbourne, VIC: Melbourne University Press.

[52] Struik, D. J. A (1987). *Concise History of Mathematics.* Mineola, NY: Dover Publications, Inc.

[53] Tufte, E. R. (1983). *The Visual Display of Quantitative Information.* Cheshire, CT: Graphics Press LLC.

[54] Tufte, E. R. (1990). *Envisioning Information.* Cheshire, CT: Graphics Press LLC.

[55] Mary-Claire van Leunen (1978). *A Handbook for Scholars.* New York, NY: Knopf.

[56] Cassio Leite Vieira (2006). *Pequeno Manual de Divulgação Científica.* Rio de Janeiro: Instituto Ciencia Hoje.

[57] Wheelock. F. M., (1992). *Wheelock's Latin Grammar.* New York, NY: HarperCollins Publishers, Inc.

[58] Yang, J. T. (1995). *An Outline of Scientific Writing: For Researchers With English as a Foreign Language.* Singapore: World Scientific.

Index

About the Authors

Marcelo Sampaio de Alencar was born in Serrita, Brazil in 1957. He received his Ph.D. from the University of Waterloo in 1994. He has more than 35 years of engineering experience, and over a quarter century as an IEEE Member, and he is currently Senior Member of the same. He has been involved with consulting and project development for many companies, agencies, and universities. He was Chair Professor at the Department of Electrical Engineering, Federal University of Campina Grande, Brazil. He is the Visiting Professor at the Department of Electrical Engineering, Federal University of Bahia, Brazil. He spent sabbatical leave working for Embratel and also for the University of Toronto, as visiting professor. He is the founder and President of the Institute for Advanced Studies in Communications (Iecom). He has been awarded several fellowships and grants. He holds an award for achievement from the College of Engineering of the Federal University of Pernambuco. He received the prestigious Attilio Giarola Medal from the Brazilian Microwave and Optoelectronic Society (SBMO). He published over 450 engineering and scientific papers. He authored 22 books and wrote chapters for ten books. Marcelo S. Alencar has contributed in different capacities to the following scientific journals: Editor of the Journal of the Brazilian Telecommunication Society (SBrT); Member of the International Editorial Board of the Journal of Communications Software and Systems (JCOMSS), published by the Croatian Communication and Information Society (CCIS); Member of the Editorial Board of the Journal of Networks (JNW), published by Academy Publisher; Founder and Editor-in-Chief of the Journal of Communication and Information Systems, jointly sponsored by the IEEE Communications Society (ComSoc) and SBrT. He is a Registered Professional Engineer and recipient of two grants from the IEEE Foundation. He is a columnist for the traditional Brazilian newspaper Jornal do Commercio, since April, 2000.

Thiago Tavares de Alencar was born in Recife, Brazil in 1985. He received his Bachelor's Degree in Law from the State University of Paraiba, in 2014. He studied at the Northdale Public School, Waterloo, Ontario, Canada, and completed his English studies, up to the level of Teacher's Training Course, at the Anglo Brazilian Institute, in Campina Grande, Brazil. He also studied Spanish and German. He taught English as a second language, to Basic and Advanced classes, at the Anglo Brazilian Institute, between 2004 and 2006. He published two papers at the 2004 IEEE Conference on the History of Electronics, Bletchley Park, England. Thiago Alencar co-autored the essay Historical Evolution of Telecommunications in Brazil, which received a grant from the IEEE Foundation.